Remote sensing of soils and vegetation in the USSR

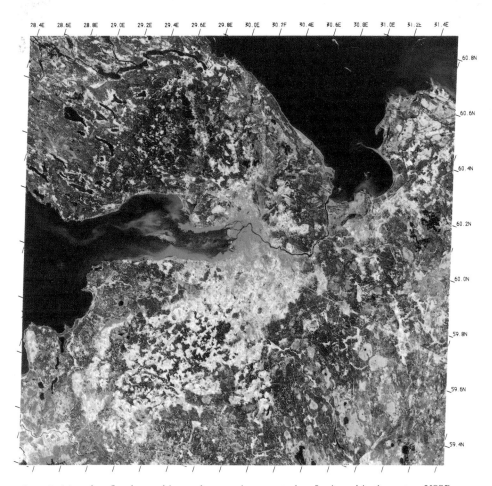

An early false-colour Landsat multispectral scanner image centred on Leningrad in the western USSR. The image covers 185 × 185 km and was recorded on 21 May 1976. The two large water bodies are the Gulf of Finland to the west and Lake Ladozhohoye to the east. The peaty soils in this region support animal grazing, forestry, potato cultivation and, to the south of Leningrad, market gardening. (Image processed by National Remote Sensing Centre, Farnborough, UK.)

Remote sensing of soils and vegetation in the USSR

P. J. Curran, G. M. Foody, K. Ya. Kondratyev,
V. V. Kozoderov and P. P. Fedchenko

Taylor & Francis
London, New York, Philadelphia

| UK | Taylor & Francis Ltd, 4 John St, London WC1N 2ET |
| USA | Taylor & Francis Inc., 1900 Frost Road, Suite 101, Bristol, PA 19007 |

Copyright © P. J. Curran, G. M. Foody, K. Ya. Kondratyev, V. V. Kozoderov and P. P. Fedchenko 1990

All rights reserved. No part of this publication may be reproduced, stored in a retrieval system, or transmitted, in any form or by any means, electronic, electrostatic, magnetic tape, mechanical, photocopying, recording or otherwise, without the prior permission of the copyright owner.

British Library Cataloguing in Publication Data

Remote sensing of soils and vegetation in the U.S.S.R.
  1. Soils. Surveying. Applications of remote sensing
  I. Curran, Paul J.
  631.4′7′028

  ISBN 0-85066-402-0

Library of Congress Cataloging in Publication Data is available

Typeset in 10/12 Baskerville by Mathematical Composition Setters Ltd, 7 Ivy Street, Salisbury, UK
Printed in Great Britain by Taylor & Francis (Printers) Ltd, Basingstoke, Hants.

Contents

Preface		*ix*
Part I	**Characterising the reflectance of soils and vegetation**	**1**
Chapter 1.	Principles of spectrometry	3
1.1.	Parameters characterising the reflectance of natural surfaces	3
1.2.	The angular distribution of radiation reflected by various natural surfaces: an example	6
1.3.	Measurements of the angular distribution of vegetation reflection from an airborne platform: an example	8
	1.3.1. The indicatrix of radiation	9
	1.3.2. The reflection-anisotropy coefficient	12
	1.3.3. The reflection-asymmetry coefficient in the Sun's vertical plane	15
1.4.	Techniques to measure the spectral reflectance of natural surfaces	16
1.5.	Instruments to measure the spectral reflectance of soils and vegetation	16
1.6.	Concluding comments on the spectral properties of natural surfaces	18
Chapter 2.	Colour and its measurement	19
2.1.	Basic definitions in colorimetry	19
2.2.	Colour mixing	20
2.3.	Basic colorimetric systems	21
2.4.	Techniques to calculate the colour and colority coordinates of non-emitting objects	24
2.5.	The ILC standard colorimetric observer	26
2.6.	Basic standards in colorimetry	26

2.7.	Colour coordinates for soils from spectral measurements acquired in the laboratory and field	31
2.8.	Determination of the number, width and range of spectral intervals used for colour calculations	35
	2.8.1. Colour coordinate calculation using a varying number of SBC values	35
	2.8.2. Colour coordinate calculation using spectral intervals of different widths	36
	2.8.3. Colour coordinate calculation using spectral intervals of different ranges	39
2.9.	Concluding comments on colour and its measurement	39

Part II Spectral reflectance of soils   41

Chapter 3.	The theory of soil reflectance	43
3.1.	A description of radiation–soil interaction using the laws of geometric optics	44
	3.1.1. The reflection of radiation by uncultivated soil	44
	3.1.2. The reflection of radiation by ploughed soil	45
	3.1.3. The reflection of radiation by harrowed soil	46
3.2.	The theory of radiation reflection from rough surfaces (tangent-plane technique)	49
	3.2.1. General solution	49
	3.2.2. Special solution of the problem for unlimited sinusoids	52
	3.2.3. Special solution of the problem for limited sinusoids	53
	3.2.4. Special solution to the problem in the case of a statistically rough surface	56
3.3.	Concluding comments on the theory of soil reflectance	62
Chapter 4.	Reflectance of soils in the laboratory and field	63
4.1.	Spectral reflectance of topsoils	63
4.2.	Spectral reflectance of soils by horizon	64
4.3.	Spectral reflectance of soil-forming rocks	70
4.4.	Spectral reflectance of soils from field measurements	70
4.5.	Spectral reflectance of soil as a function of its chemical and physical properties	73
	4.5.1. The effect of a soil's chemical properties on its spectral reflectance	75
	4.5.2. The effect of a soil's physical properties on its spectral reflectance	79
4.6.	Concluding comments on the reflectance of soils in the laboratory and field	82

Contents

Chapter 5.	Remote sensing of soil humus	83
5.1.	The reflectance properties of soil humus and its application	86
5.2.	Theoretical studies of the relationship between soil spectral reflectance and humus content	89
5.3.	Laboratory studies of the relationship between soil colour coordinates and humus content	91
5.4.	Laboratory studies on the effect of soil-forming rock on the relationship between soil colour coordinates and humus content	91
5.5.	Field studies of the relationship between colour coordinates and humus content	98
5.6.	Aircraft studies of the relationship between soil colour coordinates and humus content	99
5.7.	Concluding comments on the remote sensing of soil humus	102

PART III   Spectral reflectance of vegetation   103

Chapter 6.	Modelling of vegetation canopy reflectance	105
6.1.	Approaches to the modelling of reflectance from vegetation canopies in the USSR	105
6.2.	Three approaches to modelling the reflectance from vegetation canopies	106
6.3.	A field evaluation of reflectance models for vegetation canopies	114
6.4.	Concluding comments on modelling vegetation canopy reflectance	114
Chapter 7.	Remote sensing of crop chlorophyll	115
7.1.	The influence of species-specific leaf structure on the relationship between colour coordinates and chlorophyll concentration	117
7.2.	The relationship between colour coordinates measured in the field, chlorophyll concentration and crop yield	122
7.3.	Concluding comments on the remote sensing of crop chlorophyll	125
Chapter 8.	Remote sensing of crop state	126
8.1.	Remote sensing of winter crop state in spring	126
8.2.	Remote sensing of winter crop state in autumn	131
8.3.	Concluding comments on the remote sensing of crop state	133
Chapter 9.	Remote sensing of crop weeds	134
9.1.	Classification of weeds	134

9.2.	Weed control	135
9.3.	Conventional techniques of recording crop weediness	135
9.4.	Remote sensing of weedy crops	137
	9.4.1. Remote sensing of crop weediness in the earing phase	137
	9.4.2. Remote sensing of crop weediness during the wax-ripeness stage	145
9.5.	Concluding comments on the remote sensing of crop weeds	148

Part IV Remote sensing of soils and crops from aircraft and satellites — 149

Chapter 10.	Atmospheric correction of remotely sensed data	151
10.1.	The influence of the atmosphere on remotely sensed data	151
10.2.	Techniques for the atmospheric correction of remotely sensed data	157
10.3.	Concluding comments on the atmospheric correction of remotely sensed data	168
Chapter 11.	Remote sensing of soil and crop state	169
11.1.	Assessing the state of soils from aircraft and satellite sensor measurements	169
11.2.	Assessing the state of crops from satellite sensor measurements	173
11.3.	Concluding comments on the remote sensing of soil and crop state	180

Bibliography	181
Index	191
Author index	201

Preface

During the 1970s and 1980s publication of English- and Russian-language literature on the remote sensing of soils and vegetation increased rapidly. However, for linguistic reasons, the English-speaking research community has remained largely unaware of Soviet efforts in this field. This book should help to relieve some of this deficiency.

It was clear that a single-volume text on the use of remotely sensed data for the study of soils and vegetation in the USSR could not do justice to the vast and growing Soviet literature. For this reason we have concentrated on topics and approaches that have made Soviet work in some way different from that in the West. Therefore, the reader should be forewarned that we have avoided many topics such as land-cover classification, geographic information systems, the use of synthetic aperture radar and imaging spectrometry, and the estimation of soil moisture. Emphasis, like that of the Soviet researchers, has been placed on topics such as the use of colour as a remote-sensing transform; the interaction of electromagnetic radiation with the environment; the estimation of soil humus, leaf chlorophyll, crop weeds and crop senescence; and the scaling-up of these estimation techniques from the laboratory to the landscape.

To provide the depth of coverage needed to highlight differences that exist in this field between the English- and Russian-speaking research communities, we decided to focus on four key topics. These are the theoretical aspects of spectrometry and colour measurement; the remote sensing of soils from laboratory to aircraft; the remote sensing of vegetation from laboratory to aircraft; and the remote sensing of soils and crops from aircraft to satellite. The organisational framework and starting point for these four topics is derived, in part, from a monograph that was published in Russian by three of the authors (Kondratyev, K. Ya., Kozoderov, V. V. and Fedchenko, P. P. 1986, *Aerospace Studies of Soils and Vegetation* (Leningrad: Gidrometeoizdat)).

In compiling this book we tried to retain the spirit and detail of the original research papers and monographs. However, there was a need to take considerable editorial liberties to present the material in a form suitable for publication. Three liberties of which the reader should be aware are first, that some details that were not relevant to the method or results (e.g. on test sites and instruments), were omitted; secondly, many English terms are used with their Russian equivalents (e.g. reflectance and brightness coefficient); and thirdly, wherever reasonable, tables were replaced by graphs.

To move from an idea to the book you have before you involved the help of many. In particular, the Royal Society (UK) for supporting the initial meeting between Professor Paul Curran and Academician Professor Kirill Kondratyev; our respective institutions for their forbearance as we disappeared with piles of text and a dictionary; the National Research Council (USA) for their support of Paul Curran during the final editorial stages; the Copyright Agency of the USSR in Moscow for providing the necessary copyright clearance; Michael Dawes (Taylor and Francis, UK) for his encouragement to start this book; Dr Robin Mellors (Taylor and Francis, UK) for his encouragement to finish this book; Mrs Antonia Kostrova (Soviet Academy of Sciences, USSR) for translating a vast amount of Russian text, and Sarah Jones (Kingston Polytechnic, UK), Pat Larson, Marianne Rudolph and Dave Goude (National Aeronautics and Space Administration, USA) for their skilled assistance.

Final thanks, of course, go to our friends and families for their continued support.

PART I
Characterising the reflectance of soils and vegetation

1
Principles of spectrometry

To describe radiation interactions with soils and vegetation we need to use a set of parameters and terms. In the English and Russian literature the parameters are similar, but the terms are very different. This chapter reviews the parameters and terms used by Soviet researchers to describe the remote sensing of soils and vegetation.

1.1. Parameters characterising the reflectance of natural surfaces

Spectral reflection, the intensity of radiation reaching a sensor within a given waveband, is the key parameter upon which much optical remote sensing is based. In the Soviet remote sensing literature spectral reflection is expressed as the brightness coefficient (or reflectance) and the reflectance coefficient (or albedo) (Table 1.1). The brightness coefficient is represented by the symbol r and is defined as the ratio of the brightness or intensity of radiation, I, from a given surface in a given direction to the brightness or intensity of radiation of an ideal scatterer, I_0, under the same conditions of illumination and observation

$$r = I/I_0 \qquad (1.1)$$

As few natural surfaces are orthotropic (Lambertian) and scatter radiation equally in all directions, r varies with the viewing geometry, specifically the angles of incidence, θ_i, reflection, θ_r, and azimuth, ϕ (Figure 1.1)

$$r = r_0 f(\theta_i, \theta_r, \phi) \qquad (1.2)$$

where in this case $f = 1$.

This can be represented schematically by the so-called indicatrix of radiation. This is defined as the totality of brightness (either I or r) measured in different directions, and can be derived for different angles of incident radiation. An indicatrix of radiation can be represented in two or three dimensions (Deering

Table 1.1 Terms and units used to describe the reflectance properties of natural surfaces in the USSR.

Term	Symbol
Intensity (brightness) of radiation from a surface	I
Spectral intensity (brightness) of radiation from a surface	I_λ
Brightness coefficient (reflectance)	r
Spectral-brightness coefficient (spectral reflectance)	SBC
Reflectance coefficient (albedo)	A
Spectral-reflectance coefficient	SRC
Irradiance	E
Radiative flux	L
Radiance	R
Anisotropy coefficient	K_a

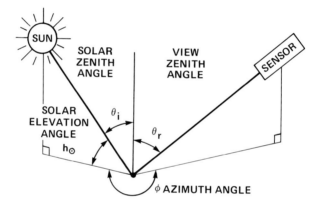

Figure 1.1 Angles used to describe the reflectance properties of a surface.

and Eck, 1987; Figure 1.2) or more simply as the difference between maximum and minimum brightness coefficients at different viewing locations.

As solar irradiance can vary over short time periods (Duggin, 1974; Slater, 1980) the brightness coefficient is, for practical purposes, usually expressed as the ratio between the brightness or intensity of radiation, I, from a given surface in a given direction, to the irradiance, E, on that surface:

$$r = I/E \tag{1.3}$$

If the hemispheric radiative flux, L_r, reflected by a surface in all directions is measured and its ratio to the hemispheric radiative flux incident on this surface, L_i, is calculated, then this ratio is called the reflectance coefficient or albedo, A:

$$A = L_r/L_i \tag{1.4}$$

Radiative fluxes L_i and L_r from Equation 1.4 are usually expressed for a unit

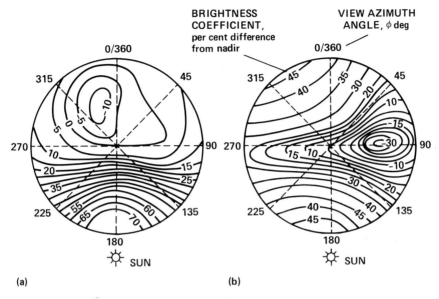

Figure 1.2 Polar brightness indicatrices for Calluna *(heather) recorded in red (0·6–0·7 μm) wavelengths: (a) mature* Calluna *and (b) mature* Calluna *with* Pteridium *(bracken). Contours are percentage difference from nadir and not all contours are shown. (Modified from Wardley et al., 1987.)*

surface, ds, by radiance, R, per unit area, s (Equation 1.5) and irradiance, E (Equation 1.6), respectively:

$$dL_i = R_s \, ds \tag{1.5}$$

$$dL_r = E \, ds \tag{1.6}$$

The reflection coefficient, A, can therefore be given as a ratio between the radiance, R_s, of a surface to its illumination, E:

$$A = R_s / E \tag{1.7}$$

A comparison of Equations 1.3 and 1.7 shows that although the reflection coefficient is a dimensionless parameter (because it is a ratio of two values of radiative flux) the brightness coefficient has dimensions which depend on I and E:

$$[r] = \frac{[I]}{[E]} = \frac{[L]}{[s][\Omega]} \frac{[s]}{[L]} = \frac{1}{[\Omega]} \tag{1.8}$$

where Ω is the solid angle. It follows that the dimensions of the brightness coefficient, r, are inversely proportional to that of the solid angle, Ω.

If it is assumed that the surface is illuminated by a parallel beam of radiation, then the brightness coefficient, r, and reflection coefficient, A, will be related:

$$A = \frac{L_r}{L_i} = r_0 \int_{\theta_r = 0}^{\pi/2} \int_{\phi = 0}^{2\pi} f(\theta_i, \theta_r, \phi) \cos \theta_r \sin \theta_r \, d\theta_r \, d\phi \tag{1.9}$$

In Soviet remote sensing research the brightness coefficient and the reflection coefficient are often used together. When the measurements are made in a specific waveband, the prefix 'spectral' is sometimes used, which leads to the abbreviation SBC for spectral-brightness coefficient and SRC for spectral-reflectance coefficient.

1.2. The angular distribution of radiation reflected by various natural surfaces: an example

Although the brightness coefficient is usually measured at nadir, some surfaces can be characterised and discriminated with greater certainty if measurements are made off-nadir. Before it is possible to define an optimum off-nadir view for determining a particular surface, the indicatrix of radiation for that surface is required. One of the first works to be published on the derivation of indicatrices of radiation was by Sharonov (1958). He concluded that although the indicatrices of radiation for natural surfaces were complex, they could be simplified into four types: specular (mirror-like), rough (very irregular), mixed (combination of specular and rough) and orthotropic (Lambertian). Since then workers have tried to quantify these four types. For example, Panfilov (1976) used

$$\gamma(\theta_i, \theta_r) = f_0 \left(1 + \frac{1 - g^2}{d[1 + g^2 - 2g \cos(\theta_i \pm \theta_r)]^{3/2}} \right) \tag{1.10}$$

where $f_0 = d/(1 + d)$, θ_i is the angle of incidence and θ_r is the angle of reflection, d is the parameter characterising the difference between specular and orthotropic (Lambertian) reflectance, and g is a measure of asymmetry in the indicatrices of radiation. Using Equation 1.10, it is possible to calculate indicatrices of radiation for some idealised cases of reflection. However, because the angular nature of reflection is complex, only a few angles of θ_i and θ_r are usually used, and certain simplifications are commonly made to minimise the known effects of light diffraction on heterogeneous surfaces (Kozoderov et al., 1980a).

A detailed discussion of both indicatrices of radiation and ratios of nadir brightness to hemispherical irradiation is presented for a range of natural surfaces in a key paper by Eaton and Dirmhirn (1979). Unfortunately, the paper does not give the reflectance characteristics of orthotropic surfaces, which hinders the transfer from values of intensity of radiation, I, from a surface to brightness coefficients, r. The paper does present a methodology for determining three facets of surface reflectance: (1) the relationship between the angular distributions of brightnesses for natural objects to the angular distribution of brightness for an ideal scatterer; (2) the relationship between the SBC and solar elevation; and (3) the effect of an altering ratio of direct to indirect solar radiation on I and SBC. This methodology was explored by Kozoderov *et al.* (1980a), to determine the relationship between the angular distributions of I for natural surfaces and those of an ideal scatterer. Indicatrices of radiation for the surfaces that Sharonov (1958) identified as being typical of the four reflectance types were used. The surfaces were snow with ice crust (specular), heterogeneous ploughed soil (rough), vegetation (mixed) and sand (orthotropic). The indicatrices of radiation allowed the calculation of the angular distribution, $I(\theta_r)$, of the upward radiance for real reflection (θ_r is the view zenith angle). Each of the four types of reflection were characterised by a definite combination of parameters d (difference between specular and Lambertian reflectance) and g (indicatrix asymmetry). As d and g were known, the angular distribution $I_\epsilon(\theta_r)$ corresponding to the reflection of an orthotropic reference (for which $g = 0$) could be calculated (Kondratyev *et al.*, 1986a).

The scheme adopted by Kozoderov *et al.* (1980a) enabled the determination of the incident diffuse radiation field. Consequently, it was possible to determine the ratio of the scattered to global radiation, where global radiation refers to the combined direct and scattered radiation. This ratio was then used to distinguish between the contributions of direct solar and global radiation to the angular distribution of reflected radiation. The relationship $r = I(\theta_r)/I_\epsilon(\theta_r)$ is the brightness coefficient for an object at different viewing angles. Estimates of r for the four surface types identified by Sharonov (1958), viewed with direct and diffuse radiation at azimuth angles (Figure 1.1) of $0°$ and $180°$, are illustrated in Figure 1.3. This shows that under direct solar irradiance the smooth snow cover exhibited specular reflection ($r > 100$ per cent) at low ($\theta_i = 60°$), medium ($\theta_i = 30°$) and high ($\theta_i = 15°$) solar elevations, the maximum reflection being greater at lower solar elevation. The angular distribution of the reflected radiation was broader than that corresponding to Fresnel reflection, with purely specular reflection. Additionally, the ice-crusted snow exhibited a non-orthotropic reflection of diffuse radiation. Because Lambertian reflection would be expected with an isotropic angular distribution and the specular reflection component should be independent of the incidence direction, this effect probably resulted from the incident radiation having a non-uniform angular distribution (Sharonov, 1958). Similar trends, but with backward reflection and of lower amplitude were present for the heterogeneous, ploughed soil (Figure 1.3).

The mixed reflection displayed maxima both in the Sun's direction and in an antisolar point. However, as the Sun's elevation above the horizon increased, the brightness coefficient declined and the amplitudes of both specular maxima increased. An almost Lambertian response was observed from the sandy surface. Here a lowering of the Sun's elevation produced a negligible decrease in the brightness coefficient. Associated with this, although not illustrated in Figure 1.3, was a simultaneous increase of the albedo, resulting from 'gliding' angles of incidence (Kondratyev and Fedchenko, 1980a).

Investigation of reflection from simple surfaces such as these enable, to a first

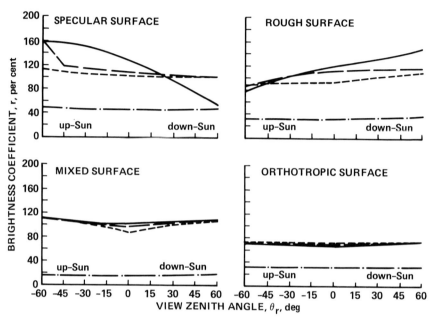

Figure 1.3 Brightness coefficients (defined here as $I(\theta_r)/I_\varepsilon(\theta_r)$) for four types of natural surface recorded at a range of solar zenith angles and view zenith angles in the solar vertical plane from $\phi = 180°$ (up-Sun) to $\phi = 0°$ (down-Sun). Note: (1) there are three solar zenith angles for direct irradiation and one composite solar zenith angle for diffuse radiation, and (2) there are nine view zenith angles from $-60°$ (up-Sun) to $60°$ (down-Sun).

approximation, an understanding of the basic laws concerning the formation of the reflected radiation field. However, general laws of solar-radiation reflection will probably require information on radiation reflection and diffraction from a surface. Consequently, the interaction between the incident radiation field and the randomly distributed scatterers on a heterogeneous surface should first be analysed.

1.3. Measurements of the angular distribution of vegetation reflection from an airborne platform: an example

Measurements of the angular spectral characteristics of reflection for various natural surfaces have been made from the Leningrad-based IL-18 'flying

Table 1.2 *Study areas in the USSR used to illustrate the reflectance properties of natural surfaces.*

	Study area	
Date	Location	Main land cover
15–16 June 1973	23 km north-east of Sverdlovsk	Coniferous forest, deciduous forest, pasture, meadow, marsh
17 June 1973	North-east Kustanai District	Mixed agricultural fields, small forest groves, small lakes
2–3 June 1975	Southwestern Yakut	Taiga
5–6 June 1975	Yakutsk region	Taiga
7 June 1975	Krasnoyarsk Krai	Taiga

laboratory' (Korzov, 1973; Kondratyev *et al.*, 1974, 1976; Korzov and Krasilschchikov, 1974; Korzov and Ter-Markaryants, 1976). Studies of the angular characteristics of vegetation reflectance have been undertaken in various regions of the USSR (Table 1.2). The aim was to derive, for selected vegetation types: (1) the indicatrices of radiation of relative (with respect to nadir) spectral brightness; (2) the indicatrices of radiation for SBCs; (3) the coefficients of anisotropy (K_a); and (4) the coefficients of reflectance asymmetry in the Sun's vertical plane (K_{sv}). These four characteristics were analysed in relation to solar elevation ($h_\odot$), viewing angles (θ_r, ϕ), wavelength (λ) and flight altitude (H_f).

1.3.1. The indicatrix of radiation

Indicatrices of radiation for vegetation had angular anisotropy which was maximised in the Sun's vertical plane ($\phi = 0°, 180°$) and minimised at azimuths perpendicular to this plane ($\phi = 90°, 270°$). This was similar to the results of other workers (e.g. Colwell, 1974). Furthermore, the SBC of vegetation was found to be positively related to the off-nadir angle, and was at its maximum when looking up-Sun.

Angular-reflection anisotropy from vegetation is wavelength-dependent, as is illustrated for a forest canopy in Figure 1.4. For instance, within the near-infrared waveband (0.96–1.38 μm) where vegetation reflectivity is high, marked variations in angular anisotropy were observed. The indicatrices of radiation at the 'hot-spot' (i.e. the point of no shadow observed looking up-Sun) was more than twice the reflection observed at nadir. However, relatively little variation was observed in the relative spectral brightness indicatrices for the forest canopy over a wide range of viewing geometries (Figure 1.4a). This is true not only in the Sun's vertical plane, but also at other azimuths. This was recorded by the constancy of the anisotropy coefficient, K_a, which averaged 1.22 ± 0.01 over the 0.96–1.38 μm spectral interval. Figure 1.4b shows that

Characterising the reflectance of soils and vegetation

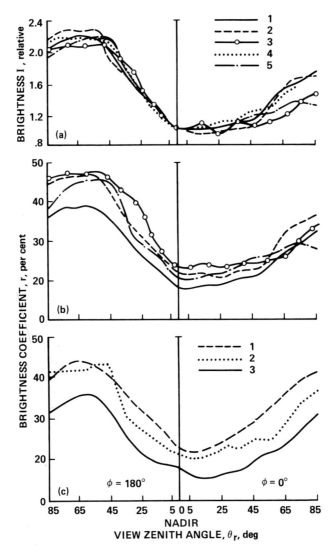

Figure 1.4 (a) Indicatrices of relative brightness, and (b) and (c) brightness coefficient for forest canopies in the solar vertical plane at $H_f = 200$ m and under different conditions of observation. (a) and (b) south-west of Yakut ASSR, 3 June 1975, $h_\odot = 31 \cdot 1°$: (1) $\lambda = 0 \cdot 96$ μm; (2) $\lambda = 0 \cdot 99$ μm; (3) $\lambda = 1 \cdot 13$ μm; (4) $\lambda = 1 \cdot 24$ μm; (5) $\lambda = 1 \cdot 38$ μm. (c) Observation regions and conditions at $\lambda = 1 \cdot 24$ μm: (1) Kustanai region, 17 June 1973, $h_\odot = 32 \cdot 5°$, 20 per cent forest canopy cover; (2) South-west Yakut ASSR, 3 June 1975, $h_\odot = 31 \cdot 1°$, 90 per cent forest canopy cover; (3) Krasnoyarsk Krai, 7 June 1975, $h_\odot = 35 \cdot 1°$, 100 per cent forest canopy cover.

Principles of spectrometry

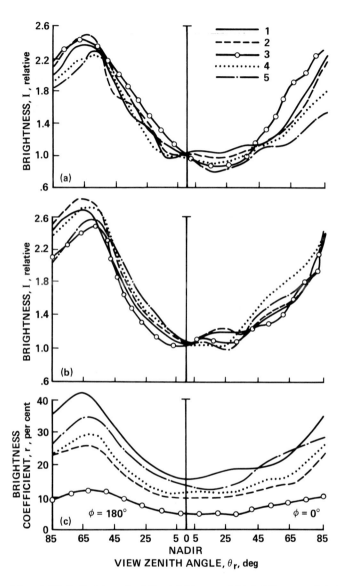

Figure 1.5 (a) and (b) Indicatrices of relative brightness and (c) brightness coefficient of a forest canopy in the solar vertical plane, obtained on 7 June 1975 over the territory of Krasnoyarsk Krai (the Upper Chadobets river) under different observation conditions. (a) $H_f = 200\ m$, $h_\odot = 35 \cdot 1°$. *(b) and (c)* $H_f = 8\ km$, $h_\odot = 28 \cdot 6°$: *(1)* $\lambda = 0 \cdot 96\ \mu m$; *(2)* $\lambda = 0 \cdot 99\ \mu m$; *(3)* $\lambda = 1 \cdot 13\ \mu m$; *(4)* $\lambda = 1 \cdot 24\ \mu m$; *(5)* $\lambda = 1 \cdot 38\ \mu m$.

the brightness, I, tended to be less variable than the reflectance, r, at all spectral intervals.

Indicatrices of radiation obtained at the same solar elevation over different forest canopies were used to compare angular-reflection anisotropy between regions. A comparison was made using the indicatrices of radiation for forest canopies measured over narrow spectral intervals which were then averaged over the $0 \cdot 96 - 1 \cdot 38$ μm spectral range. Forest canopies which, visually, appeared to be of different degrees of cover exhibited similar trends in relative indicatrices of radiation. The brightness coefficient, however, is negatively related to the degree of cover. Thus, for instance, an incomplete forest canopy was 36 per cent brighter in near-infrared wavelengths than a complete forest canopy (taiga). This was a consequence of the high reflectivity of the subforest canopy grass cover. This can be seen in Figure 1.4c, which shows the brightness coefficient indicatrices for vegetated areas containing various proportions of forest canopies at a wavelength of $1 \cdot 24$ μm. The character of the brightness coefficient's angular dependence was preserved, but the value of the brightness coefficient was a function of the proportion of the vegetation cover that is forest canopy. Thus, for instance, the brightness coefficient for a predominantly grassy surface (curve 1) was higher than that for a complete forest canopy (curve 3).

The character of vegetation angular-reflection anisotropy appeared to be independent of the flying height in near-infrared wavelengths at constant solar elevation ($h_\odot$). In absolute terms, the brightness coefficient of the vegetation cover–atmosphere system varied with altitude because of atmospheric attenuation, the effect of which was wavelength-dependent (Figure 1.5). However, as seen in Figure 1.5, the larger angular-reflection anisotropy observed at the higher altitude (8 km) was a consequence of the solar elevation being $6 \cdot 5°$ less than it was for the lower altitude (200 m).

At most viewing angles forest angular-reflection anisotropy was negatively related to the solar elevation. This was partly a consequence of the increase in the brightness coefficient at large viewing angles. Although the brightness coefficient of a forest canopy was effectively independent of solar elevation, ($h_\odot$), in every azimuthal direction over a $0-45°$ viewing angle range, significant dependence existed at larger viewing angles. For instance, at an azimuth of $180°$ and at high view zenith angles a $17°$ shift in view zenith angle could more than double the brightness coefficient of the forest canopy.

1.3.2. The reflection-anisotropy coefficient

The estimates of the reflection-anisotropy coefficient, K_a, in Figure 1.6 were derived from the spectral-brightness indicatrices of vegetation using a technique outlined by Korzov (1973). These show K_a to be dependent on the radiation wavelength and solar elevation. Over the $0 \cdot 96 - 1 \cdot 38$ μm spectral interval its relationship with wavelength was non-monotonic. However, K_a was relatively constant for the solar elevations investigated within this spectral interval, deviating by only $\pm 1 \cdot 5$ per cent from the average value of $1 \cdot 39$.

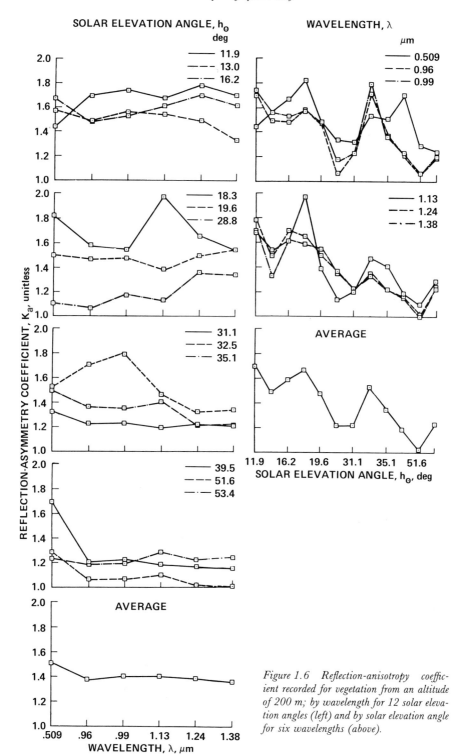

Figure 1.6 Reflection-anisotropy coefficient recorded for vegetation from an altitude of 200 m; by wavelength for 12 solar elevation angles (left) and by solar elevation angle for six wavelengths (above).

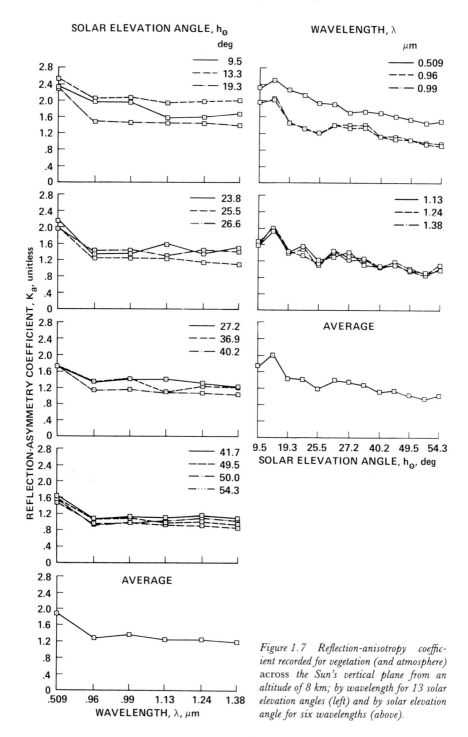

Figure 1.7 Reflection-anisotropy coefficient recorded for vegetation (and atmosphere) across the Sun's vertical plane from an altitude of 8 km; by wavelength for 13 solar elevation angles (left) and by solar elevation angle for six wavelengths (above).

Principles of spectrometry

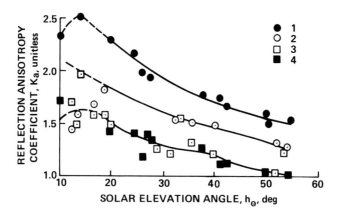

Figure 1.8 The coefficient of reflection anisotropy, K_a, for vegetation cover as a function of solar elevation angle, $h_\odot$. (1) $\lambda = 0\cdot 509$ μm, $H_f = 8000$ m; (2) $\lambda = 0\cdot 509$ μm, $H_f = 200$ m; (3) $\lambda = 0\cdot 96-1\cdot 38$ μm, $H_f = 200$ m; (4) $\lambda = 0\cdot 96-1\cdot 38$ μm, $H_f = 8000$ m.

Estimates of the reflection-anisotropy coefficients for the vegetation–atmosphere system, obtained from measurements of the spectral brightness indicatrix of a forest canopy from an altitude of 8 km, are given in Figure 1.7. A comparison of K_a averaged over the solar elevations and those observed at a single solar elevation (Figures 1.6 and 1.7) showed that vegetation–atmosphere reflection anisotropy (derived from measurements made from a height of 8 km) was larger in the visible and lower in the near-infrared wavelengths than the vegetation reflection anisotropy (derived from measurements made from a height of 200 m). The values of K_a were inversely related to the solar elevation (Figure 1.8).

At every solar elevation the reflection-anisotropy coefficient for the vegetation–atmosphere system exceeded that of vegetation in the visible wavelengths. However, in near-infrared wavelengths K_a was of a similar magnitude and displayed similar dependence on solar elevation for both the vegetation–atmosphere system and vegetation (Figure 1.8).

1.3.3. The reflection-asymmetry coefficient in the Sun's vertical plane

The reflection-asymmetry coefficient in the Sun's vertical plane, K_{SV}, was calculated using the technique described by Korzov and Ter-Markaryants (1976). Estimates show that over a wide range of solar elevations $K_{SV} < 1\cdot 0$ in near-infrared wavelengths (Figure 1.9). This was because there was a lower level of specular reflection relative to that reflected in the up-Sun direction. In visible wavelengths $K_{SV} > 1$ and was inversely related to the solar elevation.

Measurements of K_{SV}, at approximately constant solar elevation, revealed that it is inversely related to the proportion of forest cover, a relationship that is

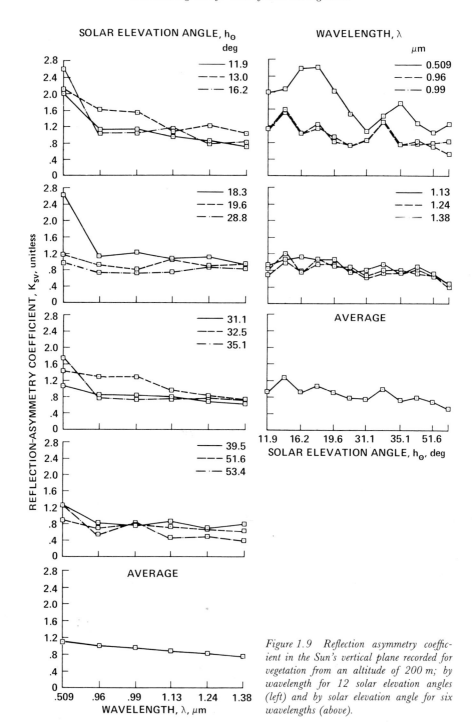

Figure 1.9 Reflection asymmetry coefficient in the Sun's vertical plane recorded for vegetation from an altitude of 200 m; by wavelength for 12 solar elevation angles (left) and by solar elevation angle for six wavelengths (above).

particularly apparent at the $0\cdot96-1\cdot38$ μm spectral interval (Figure 1.9). This resulted from the increase in the proportion of grass reflectance in the canopy reflectance, which produced an increase in the specular component of the indicatrices of radiation. However, in general the reflection coefficient in the Sun's vertical plane decreased with an increase in wavelength or, as is shown in Figure 1.9, a decrease in the solar elevation.

1.4. Techniques to measure the spectral reflectance of natural surfaces

Three techniques are commonly used to measure spectral reflectance: visual, photographic and photoelectric. These involve measuring the spectral brightness of the target surface and a reference. Visual assessments are typically derived from photometers. With the photographic techniques a reference surface is positioned in the camera's field of view along with the target, and so is recorded during the same exposure. The brightness coefficient can then be derived from the optical densities of the target and reference surface on the photograph if the film's characteristic curve is known (Curran, 1980; Lillesand and Kiefer, 1987). The photoelectric techniques, which are becoming increasingly popular in the USSR, use a sensor to measure the light flux from both the target and an ideal diffuser or reference (e.g. Milton, 1980, 1987). However, as all of these techniques are based on a comparison of the spectral brightness of the surface with that of a reference, the accuracy of the resultant SBCs will depend on optical properties of the reference surface (Feldbaum and Butkovsky, 1971) and the difference between surface and reference reflectance (Phillipson *et al.*, 1989).

1.5. Instruments to measure the spectral reflectance of soils and vegetation

The laboratory- and field-based investigations of the spectral reflectance of soils and vegetation reported in this book have used, for consistency, a limited range of apparatus. The laboratory measurements were made using an SF-18 spectrophotometer (Tarasov, 1968) in the $0\cdot4-0\cdot75$ μm spectral interval. The field measurements were made using a 'mushroom' photometer (Kondratyev and Fedchenko, 1982a; Kondratyev *et al.*, 1986a), a lens photometer and field spectrometer (Koltsov, 1975; Fedchenko and Kondratyev, 1981; Kondratyev and Fedchenko 1982b).

1.6. Concluding comments on the spectral properties of natural surfaces

Considerable research has gone into the development of remote sensing systems suitable for a wide range of applications in the USSR. Parameters ranging from the spectral, spatial, temporal and radiometric resolutions of the systems have been considered, and increasing use is being made of multi-angle measurements (similar to Kimes *et al.*, 1987) and the combinations of data from different sensing systems (Li *et al.*, 1980). However, the aim of much remote sensing research in the USSR is to use remotely sensed data to identify or classify an object (e.g. soil type or land cover) and/or to infer its characteristics (e.g. soil humus content or vegetation biomass). Consequently, the techniques for processing and manipulating the remotely sensed data need to be streamlined to achieve the goal of extracting brightness information. Necessary precursors are radiometric corrections (Teillet, 1986; Ahern *et al.*, 1987) and atmospheric corrections (Forster, 1984; Rollin *et al.*, 1985). Some aspects of these are discussed in the chapters that follow.

2
Colour and its measurement

Qualitative and quantitative characterisation of colour is a fundamental component of many remote sensing applications in the USSR. It is an expression of the spectral brightness of targets and as such is the basis of identifying objects (e.g. image classification) or estimating their parameters (Robinove, 1981; Drury, 1987; Escadafal *et al.*, 1989).

2.1. Basic definitions in colorimetry

The importance of colour as a diagnostic variable, associated with the subjective nature of its qualitative description, has led to the use of quantitative measures. Colour is an interpretation of radiation by the human eye–brain system, and has been defined as a three-dimensional parameter (Bass and Fuks, 1972). These dimensions are colour tone, λ_s, saturation, ρ_H, and brightness, I (Table 1.1). All three are required to describe an object's colour completely. The colour tone depends on the wavelength of the measured radiation, $S(\lambda)$. The saturation, or colour purity, can be derived from

$$\rho_H = \frac{I_\lambda}{I_\lambda + I_\sigma} \qquad (2.1)$$

where I_λ is the spectral brightness of the object and I_σ is the brightness of a white reference body. Colour tone and saturation combined allow a qualitative description of colour. A quantitative description requires these in addition to the brightness of the object. Brightness, I, is in direct proportion to the luminous power, dIf, of the surface element, ds, in a given direction and in inverse proportion to the area of its projection onto the plane perpendicular to the direction of the light propagation (Brandt and Tageeva, 1967). It can be derived from

$$I = \frac{dIf}{ds \cos \phi_0} \qquad (2.2)$$

where ϕ_0 is the azimuth angle between the direction of light propagation and the illuminated element, ds.

2.2. Colour mixing

The investigation of colour mixing has a long history. The English scientist Sir Isaac Newton showed that if all the colours of the spectrum are mixed together, white is the result (Committee on Colorimetry, 1963). Additionally, the mixing of two or more spectral colours will produce a new colour, although it may not be a spectral colour.

Contemporary colorimetry is based on three laws, proposed by Grassmann in the 19th century (Bouma, 1971; MacAdam, 1985). These are:

(1) Any four colours are linearly interdependent, but the number of linearly independent three-colour systems is limitless.
(2) If in a mixture of three colours one of them is changed, so too is the colour of the mixture.
(3) The colour of the mixture is determined only by the colours of the light producing it. It is independent of the spectral composition of the radiation producing the lights to be mixed.

The 'linearity laws' of colour (Dzhad and Vyshetski, 1978) arise as a consequence of the last of these laws. These can be envisaged in terms of the following examples.

(1) If the colour A is the same as the colour B and the colour C is the same as the colour D, then the colour of the mixture $A + C$ is identical to that of $B + D$.
(2) If the colour A is the same as the colour B and the three-colour mixture C identical to the mixture D, then, upon subtracting A from the mixture C the remaining colour will be the same as that obtained by subtracting B from D.
(3) If the radiation fluxes for each colour in a mixture are changed by the same quantity, the colour of the mixture will remain constant.

Any colour can be obtained by mixing, in the correct proportions, three mutually independent colours: red R (0·700 µm), green G (0·546 µm) and blue B (0·436 µm) (Chamberlin, 1951). These three colours can be combined to express colour C as

$$C = rR + gG + bB \qquad (2.3)$$

where R, G and B are unit quantities of the basic colours and r, g and b represent the weights which express the proportion of red, green and blue colours in a given mixture. Consequently, for white the colour equation would be

$$C = \tfrac{1}{3}(R + G + B) \tag{2.4}$$

Sometimes the Soviet literature refers to the R, G and B coefficients as colour coordinates and the r, g and b coefficients as colority coordinates, although different terminology can be found in the general literature on colour (e.g. Judd and Wyszecki, 1975; Kelly and Judd, 1976; Hunt, 1987), that adopted in the Soviet literature (e.g. Gurevich, 1950, 1968; Kondratyev et al., 1986a) is used here. The latter are used for the qualitative description of colour and can be determined from

$$r = \frac{R}{R + G + B} \qquad g = \frac{G}{R + G + B} \qquad b = \frac{B}{R + G + B} \tag{2.5}$$

where

$$r + g + b = 1 \tag{2.6}$$

2.3. Basic colorimetric systems

Three basic systems are used for expressing colour, and these are commonly referred to as $B_{\lambda p}$, RGB and XYZ (MacAdam, 1985).

The $B_{\lambda p}$ system is relatively simple and provides purple as well as the spectral colours. Its basic principle is that any colour can be produced by mixing white with a colour. The parameters characterising colority in this system are the colour purity (ρ_H) and colour tone (λ_s). The former is an expression of the portion of a pure spectral colour in a given mixture. The latter is determined by the wavelength of the spectral colour which must be mixed with white to give the desired mixture. This colorimetric system does not, however, allow colour calculations, and so it is used little in Soviet remote sensing research.

The RGB colour system arose from experiments by Wright and Guild (Wright, 1969). It is a simple system which allows the solution of many colorimetric problems as well as colour quantification and colour calculations. The RGB system can be drawn schematically as the colour triangle (Figure 2.1). The wavelengths $\lambda_r = 0 \cdot 700$ μm, $\lambda_g = 0 \cdot 546$ μm and $\lambda_b = 0 \cdot 436$ μm were selected by the International Luminance Council (ILC) and the Commission International de l'Éclairage (CIE) as those representing monochromatic emissions of red, green and blue colours, respectively. These three basic colours form the apexes of the colour triangle with white at its centre. Inside the triangle lie the colours obtained from the mixture of the three basic colours; others lie outside. The position of the mixture within the triangle depends on the relative proportions of the basic colours. However, the use of this system is complicated by the possibility of obtaining negative colour coordinates which hinder colour calculations.

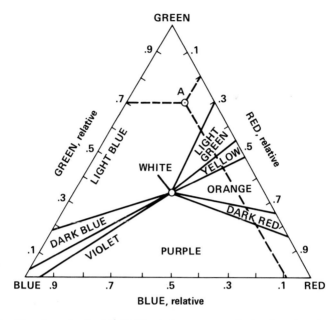

Figure 2.1 The colour triangle of the RGB *colorimetric system. Broken lines show the location of any point of the colour triangle (e.g. A) from colour coordinates.*

The *XYZ* system also uses the basic colours, red, green and blue (Hunt, 1987). With this system there are three requirements.

(1) The existing colours are all inside the colour triangle, and hence all coordinates are positive.
(2) The quantitative estimate of colour is determined from the *Y*-coordinate; the light flux for this coordinate corresponds to $0 \cdot 680$ μm. Zero brightness corresponds to *X* and *Z*.
(3) To obtain white the colority coordinates must satisfy the equation $x = y = z = \frac{1}{3}$.

The basic colours of the *XYZ* system can be expressed in terms of the *RGB* system as

$$X = a_1 R + a_2 G + a_3 B$$
$$Y = a_4 R + a_5 G + a_6 B$$
$$Z = a_7 R + a_8 G + a_9 B$$

where a_1 to a_9 are constants (Hunt, 1987) and the colour *C* can be expressed in the *XYZ* system as

$$C = xX + yY + zZ \qquad (2.7)$$

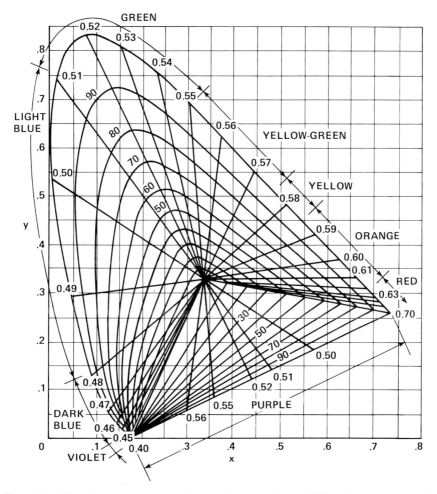

Figure 2.2 The colour graph in rectangular coordinates used in the XYZ colorimetry system. The figures near the edge of the graph are wavelengths in μm.

The colority coordinates and colour coordinates can be related by (Hunt, 1987)

$$x = \frac{X}{X+Y+Z} \quad y = \frac{Y}{X+Y+Z} \quad z = \frac{Z}{X+Y+Z} \quad (2.8)$$

with $x + y + z = 1$.

As with the *RGB* system, only two of the coordinates in the *XYZ* system are independent, the third ensuring that their sum is unity. Consequently, the determination of any colour requires only two values, and, instead of a triangle, a system of rectangular coordinates may be used (Figure 2.2). Inside the curve of monochromatic colours are the real colours. The *XYZ* system is widely used for colour calculation in the USSR, and was used for most of the following calculations of colour characteristics.

2.4. Techniques to calculate the colour and colority coordinates of non-emitting objects

Two commonly used techniques for calculating colour and colority coordinates are the methods of weighted ordinates and selected ordinates (Hunter and Harold, 1987). Both require knowledge of the relative spectral distribution of the radiative flux incident on the target, the target's SRC (Table 1.1) and the adding functions.

To derive the colour coordinates from the weighted-ordinate method, it is assumed that the entire spectrum is divided into narrow intervals $\Delta\lambda$ (Judd and Weyszecki, 1975). Furthermore, the radiation within each interval is assumed constant. Using the colorimetric system adopted by the ILC in 1931, the XYZ colour coordinates can be derived from (Judd and Wyszecki, 1975; Escadafal et al., 1989)

$$X = k \sum_{i=1}^{n} S(\lambda) \, r(\lambda) \, \bar{x}(\lambda) \, \Delta\lambda$$

$$Y = k \sum_{i=1}^{n} S(\lambda) \, r(\lambda) \, \bar{y}(\lambda) \, \Delta\lambda \qquad (2.9)$$

$$Z = k \sum_{i=1}^{n} S(\lambda) \, r(\lambda) \, \bar{z}(\lambda) \, \Delta\lambda$$

where

$$k = 100 \bigg/ \sum_{i=1}^{n} S(\lambda) \, \bar{y}(\lambda) \, \Delta\lambda$$

and $S(\lambda)$ is the relative spectral distribution of the incident radiation in the spectral region $\Delta\lambda$, n is the number of spectral intervals, $r(\lambda)$ is the target's spectral-reflectance coefficient and $\bar{x}(\lambda)$, $\bar{y}(\lambda)$ and $\bar{z}(\lambda)$ are the adding functions, with

$$x(\lambda) = \frac{\bar{x}(\lambda)}{\bar{x}(\lambda) + \bar{y}(\lambda) + \bar{z}(\lambda)}$$

$$y(\lambda) = \frac{\bar{y}(\lambda)}{\bar{x}(\lambda) + \bar{y}(\lambda) + \bar{z}(\lambda)} \qquad (2.10)$$

$$z(\lambda) = \frac{\bar{z}(\lambda)}{\bar{x}(\lambda) + \bar{y}(\lambda) + \bar{z}(\lambda)}$$

and

$$x(\lambda) + y(\lambda) + z(\lambda) = 1$$

With the selected-ordinate technique, however, the spectrum is divided into spectral intervals of different widths. The latter were determined by the spectral distribution of radiation, $S(\lambda)$, and by the adding functions.

Table 2.1 Selected ordinates (λ, μm) used to calculate the colour coordinates of a non-emitting object with the 1931 ILC system, with respect to standard radiation A (see Section 2.6.)

Number of the ordinate	Colour coordinates		
	X	Y	Z
1	0·4440	0·4878	0·4164
2*	0·5169	0·5077	0·4249
3	0·5440	0·5173	0·4294
4	0·5542	0·5241	0·4329
5*	0·5614	0·5298	0·4360
6	0·5671	0·5348	0·4387
7	0·5720	0·5394	0·4413
8*	0·5763	0·5437	0·4437
9	0·5802	0·5478	0·4460
10	0·5839	0·5517	0·4483
11*	0·5872	0·5554	0·4505
12	0·5905	0·5591	0·4526
13	0·5935	0·5627	0·4547
14*	0·5965	0·5663	0·4568
15	0·5994	0·5698	0·4588
16	0·6023	0·5733	0·4608
17*	0·6052	0·5769	0·4629
18	0·6080	0·5805	0·4649
19	0·6109	0·5841	0·4670
20*	0·6138	0·5879	0·4692
21	0·6169	0·5918	0·4716
22	0·6200	0·5959	0·4741
23*	0·6233	0·6001	0·4768
24	0·6269	0·6047	0·4799
25	0·6308	0·6097	0·4834
26*	0·6353	0·6152	0·4875
27	0·6405	0·6215	0·4927
28	0·6469	0·6292	0·4993
29*	0·6559	0·6397	0·5084
30	0·6735	0·6590	0·5267
Weighting coefficients			
for 30 ordinates	0·003 661	0·003 333	0·001 185
for 10 ordinates	0·010 984	0·01	0·003 555

NB: The ordinates marked with an asterisk are used in simplified calculations.

For standard emissions the selected ordinates are determined for the cases when the spectrum is divided into 10, 30 and 100 ordinates (Table 2.1). Because this technique is less accurate than that of the weighted ordinates, the following calculations were made using the weighted-ordinate method.

Further information on the concepts and techniques mentioned is beyond the scope of this book, and may be found in the general literature on colour and its measurement (e.g. Bouma, 1971; Judd and Wyszecki, 1975; Hunt, 1987; Hunter and Harold, 1987). A remote sensing application of some of these measures is given by Escadafal et al. (1989).

2.5. The ILC standard colorimetric observer

In 1931 the ILC recommended a standard colorimetric observer (SCO) (based on a $2°$ field-of-view) for the ready estimation of colour characteristics (Table 2.2). In 1964 the ILC also recommended a supplementary SCO based on a $10°$ field-of-view (Hunter and Harold, 1987). The SCO could be used to determine the colour coordinates for both emitting and non-emitting objects. The SCO for colour calculations is usually presented as tables of coefficients, referred to as specific colour coordinates or adding functions (Tables 2.2 and 2.3) that vary with the size of the angular field over which the measurements are made (Figure 2.3) (Merik, 1968; Judd and Wyszecki, 1975).

2.6. Basic standards in colorimetry

To calculate the colour coordinates for a non-emitting object, three parameters must be known. These are (1) the relative spectral distribution of the radiation incident on the object, (2) the adding functions and (3) the visible SRC or SBC of the object.

For colorimetric calculations the ILC recommended standardised radiation sources. Initially the ILC recommended three sources: A, B and C. Source A displays a radiative flux with the same spectral distribution as that from an absolute blackbody with a colour temperature of approximately 2856 K (Le Grand, 1968); where the precise value depends on the magnitude of one of the constants used in the formula describing Planck's law (Hunt, 1987). Source B reproduces direct solar radiation, with the spectral composition of an absolute blackbody at a colour temperature of approximately 4870 K. Source C simulates daylight with a colour temperature of approximately 6770 K (Hunt, 1987). As emissions from sources B and C do not have the same spectral composition as daylight (Dzhad and Vyshetski, 1978). Additional sources have been proposed. The standard emission source D_{65}, for instance, has a spectral composition similar to daylight and a colour temperature of approximately 6500 K. However,

Table 2.2 The 1931 ILC standard colorimetric observer (SCO)

Wavelength (μm)	Colour coordinates			Wavelength (μm)	Colour coordinates		
	$\bar{x}(\lambda)$	$\bar{y}(\lambda)$	$\bar{z}(\lambda)$		$\bar{x}(\lambda)$	$\bar{y}(\lambda)$	$\bar{z}(\lambda)$
0·380	0·0014	0·0000	0·0065	0·585	0·9786	0·8163	0·0014
0·385	0·0022	0·0001	0·0105	0·590	1·0263	0·7570	0·0011
0·390	0·0042	0·0001	0·0201	0·595	1·0567	0·6949	0·0010
0·395	0·0076	0·0002	0·0362	0·600	1·0622	0·6310	0·0008
0·400	0·0143	0·0004	0·0679	0·605	1·0456	0·5668	0·0006
0·405	0·0232	0·0006	0·1102	0·610	1·0026	0·5030	0·0003
0·410	0·0435	0·0012	0·2074	0·615	0·9384	0·4412	0·0002
0·415	0·0776	0·0022	0·3713	0·620	0·8544	0·3810	0·0002
0·420	0·1344	0·0040	0·6456	0·625	0·7514	0·3210	0·0001
0·425	0·2148	0·0073	1·0391	0·630	0·6424	0·2650	0·0000
0·430	0·2839	0·0116	1·3856	0·635	0·5419	0·2170	0·0000
0·435	0·3285	0·0168	1·6230	0·640	0·4479	0·1750	0·0000
0·440	0·3483	0·0230	1·7471	0·645	0·3608	0·1382	0·0000
0·445	0·3481	0·0298	1·7826	0·650	0·2835	0·1070	0·0000
0·450	0·3362	0·0380	1·7721	0·655	0·2187	0·0816	0·0000
0·455	0·3187	0·0480	1·7441	0·660	0·1649	0·0610	0·0000
0·460	0·2908	0·0600	1·6692	0·665	0·1212	0·0446	0·0000
0·465	0·2511	0·0739	1·5281	0·670	0·0874	0·0320	0·0000
0·470	0·1954	0·0910	1·2876	0·675	0·0636	0·0232	0·0000
0·475	0·1421	0·1126	1·0419	0·680	0·0468	0·0170	0·0000
0·480	0·0956	0·1396	0·8130	0·685	0·0329	0·0119	0·0000
0·485	0·0580	0·1693	0·6162	0·690	0·0227	0·0082	0·0000
0·490	0·0320	0·2080	0·4652	0·695	0·0158	0·0057	0·0000
0·495	0·0147	0·2586	0·3533	0·700	0·0114	0·0041	0·0000
0·500	0·0049	0·3230	0·2720	0·705	0·0081	0·0029	0·0000
0·505	0·0024	0·4073	0·2123	0·710	0·0058	0·0021	0·0000
0·510	0·0093	0·5030	0·1582	0·715	0·0041	0·0015	0·0000
0·515	0·0291	0·6082	0·1117	0·720	0·0029	0·0010	0·0000
0·520	0·0633	0·7100	0·0782	0·725	0·0020	0·0007	0·0000
0·525	0·1096	0·7932	0·0573	0·730	0·0014	0·0005	0·0000
0·530	0·1655	0·8620	0·0422	0·735	0·0010	0·0004	0·0000
0·535	0·2257	0·9149	0·0298	0·740	0·0007	0·0002	0·0000
0·540	0·2904	0·9540	0·0203	0·745	0·0005	0·0002	0·0000
0·545	0·3597	0·9803	0·0134	0·750	0·0003	0·0001	0·0000
0·550	0·4334	0·9950	0·0087	0·755	0·0002	0·0001	0·0000
0·555	0·5121	10·0000	0·0057	0·760	0·0002	0·0001	0·0000
0·560	0·5945	0·9950	0·0039	0·765	0·0001	0·0000	0·0000
0·565	0·6784	0·9786	0·0027	0·770	0·0001	0·0000	0·0000
0·570	0·7621	0·9520	0·0021	0·775	0·0001	0·0000	0·0000
0·575	0·8425	0·9154	0·0018	0·780	0·0000	0·0000	0·0000
0·580	0·9163	0·8700	0·0017	Total	21·3714	21·3711	21·3715

Table 2.3 The additional 1964 ILC standard colorimetric observer.

Wavelength (μm)	Colour coordinates			Wavelength (μm)	Colour coordinates		
	$\bar{x}_{10}(\lambda)$	$\bar{y}_{10}(\lambda)$	$\bar{z}_{10}(\lambda)$		$\bar{x}_{10}(\lambda)$	$\bar{y}_{10}(\lambda)$	$\bar{z}_{10}(\lambda)$
0·380	0·0002	0·0000	0·0007	0·585	1·0743	0·8256	0·0000
0·385	0·0007	0·0001	0·0029	0·590	1·1185	0·7774	0·0000
0·390	0·0024	0·0003	0·0105	0·595	1·1343	0·7204	0·0000
0·395	0·0072	0·0008	0·0323	0·600	1·1240	0·6583	0·0000
0·400	0·0191	0·0020	0·0860	0·605	1·0891	0·5939	0·0000
0·405	0·0434	0·0045	0·1971	0·610	1·0305	0·5280	0·0000
0·410	0·0847	0·0088	0·3894	0·615	0·9507	0·4618	0·0000
0·415	0·1406	0·0145	0·6568	0·620	0·8563	0·3981	0·0000
0·420	0·2045	0·0214	0·9725	0·625	0·7549	0·3396	0·0000
0·425	0·2647	0·0295	1·2825	0·630	0·6475	0·2835	0·0000
0·430	0·3147	0·0387	1·5535	0·635	0·5351	0·2283	0·0000
0·435	0·3577	0·0496	1·7985	0·640	0·4316	0·1798	0·0000
0·440	0·3837	0·0621	1·9673	0·645	0·3437	0·1402	0·0000
0·445	0·3867	0·0747	2·0273	0·650	0·2683	1·1076	0·0000
0·450	0·3707	0·0895	1·9948	0·655	0·2043	0·0812	0·0000
0·455	0·3430	0·1063	1·9007	0·660	0·1526	0·0603	0·0000
0·460	0·3023	0·1282	1·7454	0·665	0·1122	0·0441	0·0000
0·465	0·2541	0·1528	1·5549	0·670	0·0813	0·0318	0·0000
0·470	0·1956	0·1852	1·3176	0·675	0·0579	0·0226	0·0000
0·475	0·1323	0·2199	1·0302	0·680	0·0409	0·0159	0·0000
0·480	0·0805	0·2536	0·7721	0·685	0·0286	0·0111	0·0000
0·485	0·0411	0·2977	0·5701	0·690	0·0199	0·0077	0·0000
0·490	0·0162	0·3391	0·4153	0·695	0·0138	0·0054	0·0000
0·495	0·0051	0·3954	0·3024	0·700	0·0096	0·0037	0·0000
0·500	0·0038	0·4806	0·2185	0·705	0·0066	0·0026	0·0000
0·505	0·0154	0·5314	0·1592	0·710	0·0046	0·0018	0·0000
0·510	0·0375	0·6067	0·1120	0·715	0·0031	0·0012	0·0000
0·515	0·0714	0·6857	0·0822	0·720	0·0022	0·0008	0·0000
0·520	0·1177	0·7618	0·0607	0·725	0·0015	0·0006	0·0000
0·525	0·1730	0·8233	0·0143	0·730	0·0010	0·0004	0·0000
0·530	0·2365	0·8752	0·0305	0·735	0·0007	0·0003	0·0000
0·535	0·3042	0·9238	0·0206	0·740	0·0005	0·0002	0·0000
0·540	0·3768	0·9620	0·0137	0·745	0·0004	0·0001	0·0000
0·545	0·4516	0·9822	0·0079	0·750	0·0003	0·0001	0·0000
0·550	0·5298	0·9918	0·0040	0·755	0·0002	0·0001	0·0000
0·555	0·6161	0·9991	0·0011	0·760	0·0001	0·0000	0·0000
0·560	0·7052	0·9973	0·0000	0·765	0·0001	0·0000	0·0000
0·565	0·7938	0·9824	0·0000	0·770	0·0001	0·0000	0·0000
0·570	0·8787	0·9556	0·0000	0·775	0·0001	0·0000	0·0000
0·575	0·9512	0·9152	0·0000	0·780	0·0000	0·0000	0·0000
0·580	1·0142	0·8689	0·0000	Total	23·3294	23·3324	23·3343

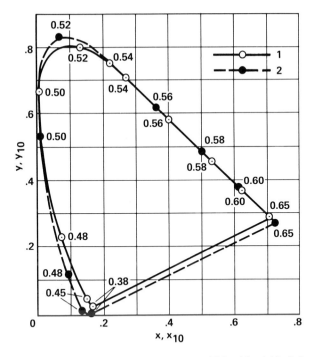

Figure 2.3 A comparison of ILC colour graphs for (1) 1964 SCO with a field-of-view of 10° and (2) 1931 SCO with a field-of-view of 2°. The figures near points are wavelengths in μm.

the spectral composition of daylight is modulated by a variety of factors, including the effect of clouds, diurnal and seasonal variation in spectral irradiance and the relative proportions of direct and diffuse light. Such factors must be borne in mind when attempting to measure the colour of an object *in situ*, especially of large natural objects such as forests, lakes or fields.

The adding functions were the second parameter listed as necessary for calculating colour coordinates. These can usually be derived from prepared tables. In colorimetric calculations the tables used are often presented as products of the adding functions with the relative spectral distribution of energy from standard emission sources.

The wavelength dependence of the adding functions is shown in Figure 2.4. This also illustrates that the adding coefficients of the SCOs measured over different angular fields are dissimilar. This indicates an imperfect estimation of the adding functions may be made, which could decrease the accuracy of the resultant colour coordinate calculations.

The final parameter needed to accomplish colorimetric calculations is the SRC measured in visible wavelengths. A detailed discussion of this parameter's variability is given below. However, it is essential to note that the accuracy of the SRC measurements is largely determined by the quality of the reference surface used in the calculation. Various surfaces can be used; some commonly used in

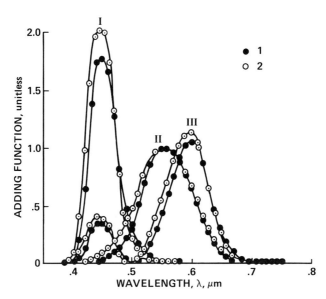

Figure 2.4 A comparison of (1) ILC 1931 SCO observer adding functions (2° field-of-view) with (2) the adding functions of the additional ILC 1964 SCO (10° field-of-view). Adding functions $III = \bar{x}(\lambda), \bar{x}_{10}(\lambda); II = \bar{y}(\lambda), \bar{y}_{10}(\lambda); I = \bar{z}(\lambda), \bar{z}_{10}(\lambda)$.

the USSR are barium sulphate, magnesium oxide, barite paper, ground glass and grey cards. None is completely suited to laboratory and field studies of SRC, because they are neither spectrally neutral nor Lambertian reflectors (Kimes and Kirchner, 1982; Milton, 1987, 1989; Leshkevich, 1988).

Finally, the ILC has recommended four standard conditions of illumination and observation for determining the colorimetric characteristics of opaque objects observed in remote sensing studies (Judd and Wyszecki, 1975; Dzhad and Vyshetski, 1978).

(1) 0/45 (Figure 2.5a). The object is illuminated by a light beam whose axis forms an angle with the normal to the object not larger than 10° (solar zenith angle, Figure 1.1). The object is observed at an angle of 45 ± 5° with respect to the normal (view zenith angle, Figure 1.1). For both the illuminating and observing beams the angle between the beam axis and any of its rays must not exceed 5°.

(2) 45/0 (Figure 2.5b). The object is illuminated by one or more light beams, the axes of which form an angle of 45 ± 5° with respect to the normal to the object's surface (solar zenith angle, Figure 1.1). The angle between the normal and the view direction (view zenith angle, Figure 1.1) must not exceed 10°. For both the illuminating and the observed beams, the angle between the axis of the beam and any of its rays must not exceed 5°.

(3) 0/Diff. (Figure 2.5c). The object is illuminated by a light beam whose axis forms an angle with the normal to the sample not larger than 10° (solar

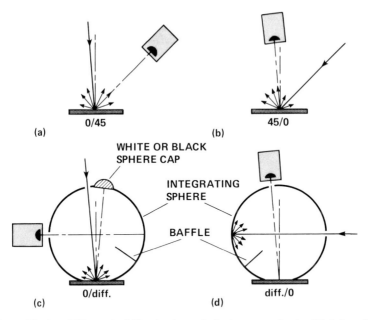

Figure 2.5 The four CIE standard illuminating and viewing geometries (modified from Judd and Wyszecki, 1975).

zenith angle, Figure 1.1). The angle between the axis of the illuminating beam and any of its rays must not exceed 5°. The reflected flux is collected using an integrating sphere of any diameter on the condition that the total aperture area does not exceed 10 per cent of the inner reflecting surface of the sphere.

(4) Diff./0 (Figure 2.5d). The object is illuminated with diffuse light using an integrating sphere. The angle between the normal to the sample and the view direction (view zenith angle, Figure 1.1) must not exceed 10°. The angle between the axis of the observed light beam and any of its rays must not exceed 5°. The integrating sphere may be of any diameter as long as the total area of apertures does not exceed 10 per cent of the sphere's inner reflecting surface.

2.7. Colour coordinates for soils from spectral measurements acquired in the laboratory and field

To determine an object's colour coordinates it is necessary to know its spectral reflectance as measured by the SBC and the spectral composition of the incident radiation. The former can be derived from data recorded by a variety of remote sensors. Because the spectral composition of the commonly available D_{65} emission source closely corresponds to that of global radiation, this source was used in the studies discussed below.

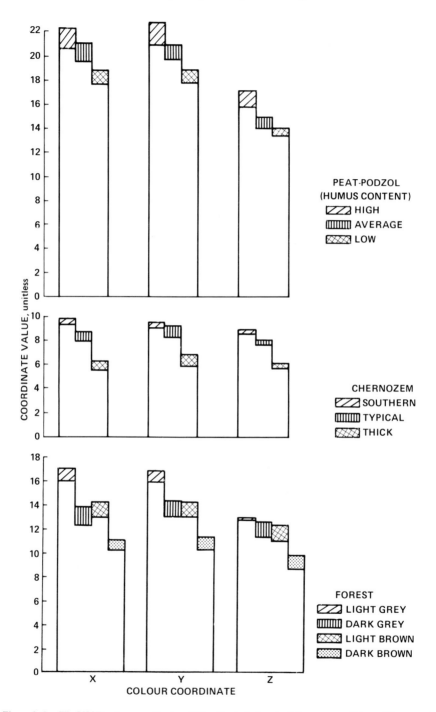

Figure 2.6 The XYZ colour coordinates of 10 soils studied under laboratory conditions. The range is the 70 per cent confidence limit calculated from five colour measurements.

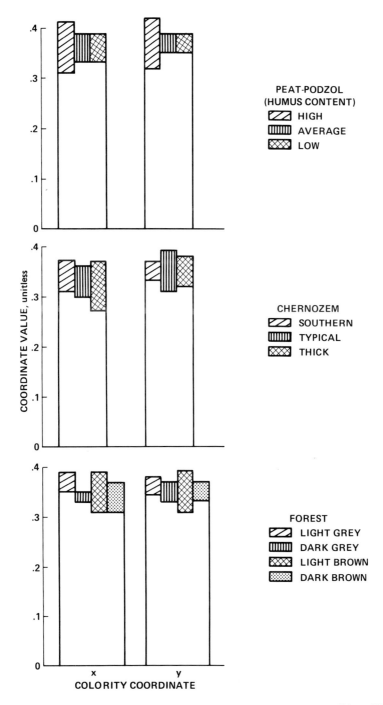

Figure 2.7 The XYZ colority coordinates of 10 soils studied under laboratory conditions. The range is the 70 per cent confidence limit calculated from five colority measurements.

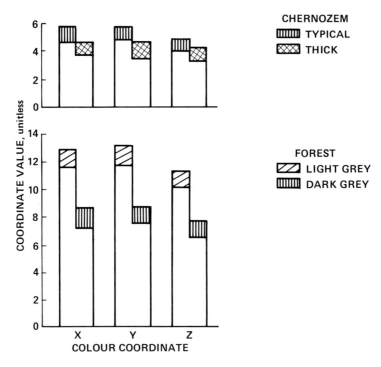

Figure 2.8 The XYZ *colour coordinates of four soils studied under field conditions. The range is the 70 per cent confidence limit calculated from five colour measurements.*

Soil samples from the upper horizons of arable soils in the Ukraine and Kaluga regions of the USSR were analysed under laboratory conditions. To remove the effects of soil structure and moisture on spectral reflectance, the samples were dried and sieved. The spectral reflectance of each sample was then measured with an SF-18 spectrophotometer in the $0\cdot40-0\cdot75\mu m$ spectral region.

The calculated colour coordinates differed greatly between the soil types (Figure 2.6). Peat–podzol soils typically displayed the largest colour coordinates. The light and dark grey forest soils had colour coordinates 15–30 per cent lower than the peat–podzol soils. The lowest colour coordinates were observed for the chernozem soils with high humus contents. Consequently, there appears to be a negative relationship between the colour coordinates and the humus content. This is discussed in greater detail in Chapters 4 and 5.

The colority coordinates for these soils, derived from Equation 2.10, showed greater between-soil-type overlap (Figure 2.7). As a consequence the colour coordinates seem to be a more reliable diagnostic variable than the colority coordinates.

A quantitative estimate of soil colour was also derived under fielde conditions in the Ukraine region of the USSR using data from an airborne sensor (Figure

2.8). The values of soil colour were found to be 1·5–2·0 times lower than those derived in the laboratory (Figure 2.6). This was a result of the more complex field environment where it is not always possible to allow for the various factors affecting spectral reflectance (Tolchelnikov, 1974; Kondratyev and Fedchenko, 1980b,c,d; Fedchenko, 1981; Fedchenko and Kondratyev, 1981; Rachkulik and Sitnikova, 1981).

2.8. Determination of the number, width and range of spectral intervals used for colour calculations

Quantitative estimation of the colour of objects is derived from the measurements of SBC and SRC in visible and near-visible wavelengths. Because it is laborious to estimate the SBC at a high spectral resolution (e.g. 10 nm is often recommended), it is desirable to identify a procedure which enables the accurate determination of the colour coordinates at a reduced spectral resolution. Two approaches are discussed. First, the SBC of the target is measured for only a limited number of spectral intervals and the spectral-reflectance curve is generated by extrapolation and interpolation. Values of SBC, obtained in 10, 20, 40 or 50 nm steps, are used to calculate colour coordinates. In the second approach the SBC is measured in 10, 20, 40 or 50 nm steps, and the colour coordinates are calculated directly from these measurements only. Finally, the sensitivity of the colour estimate to spectral range is considered.

2.8.1. Colour coordinate calculation using a varying number of SBC values

Continuous reflectance curves for soils and vegetation were obtained using an SF-18 spectrophotometer. From these curves the SBC values obtained every 10 nm were used to calculate the colour coordinates X, Y and Z. The same reflectance curves were then sampled at only a few selected points. The SBC values were obtained at wavelengths of 0·44, 0·54, 0·64 and 0·74 μm for soil (Condit, 1970) and at wavelengths of 0·50, 0·55, 0·60, 0·65 and 0·75μm for vegetation (Kharin, 1975). From these data a complete reflectance curve was generated by extrapolation and interpolation from these points. The SBCs were derived from this curve at 10 nm intervals and the colour coordinates X, Y and Z were calculated. This procedure was repeated after dropping the observation of reflectance at a wavelength of 0·74 μm for soil and 0·75 μm for vegetation. Consequently, the SBC had been estimated at three different spectral resolutions: 41, 4 and 5 or 3 wavebands, where 41 is for the 'continuous' reflectance curve.

The differences in the calculated colour coordinates from the three different approaches were negligible (Figure 2.9). Consequently, the colour coordinates of soils and vegetation may be adequately estimated from a sample of three carefully selected wavelengths from which the complete reflectance curve can then be constructed. This allows the use of simple measuring instruments.

Characterising the reflectance of soils and vegetation

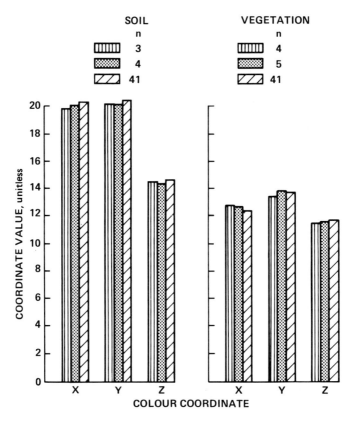

Figure 2.9 Results of colour coordinate calculations using values from the spectral reflectance curve for vegetation and soils. Samples of 3, 4 and 41 values were used for soils and samples of 4, 5 and 41 values were used for vegetation.

However, errors can be introduced because of a lack of *a priori* information on the shape of the reflectance curve. The wavelengths or sample points therefore need to be carefully selected if this technique is to be successful.

2.8.2. Colour coordinate calculation using spectral intervals of different widths

Colour coordinates were calculated from complete spectral-reflectance curves of crops and soils derived from measurements with an SF-18 spectrophotometer. These were then sampled every 10, 20, 40 and 50 nm and the SBC was estimated for each interval size. The estimated SBCs (Figure 2.10) were similar for each interval size. Thus, the eight values of SBC measured at wavelengths of $0 \cdot 40$, $0 \cdot 45$, $0 \cdot 50$, $0 \cdot 55$, $0 \cdot 60$, $0 \cdot 65$, $0 \cdot 70$ and $0 \cdot 75$ μm were sufficient for an accurate calculation of the colour coordinates for these crops and soils, as the error in SBC estimate was less than 5 per cent for the largest (50 nm) interval.

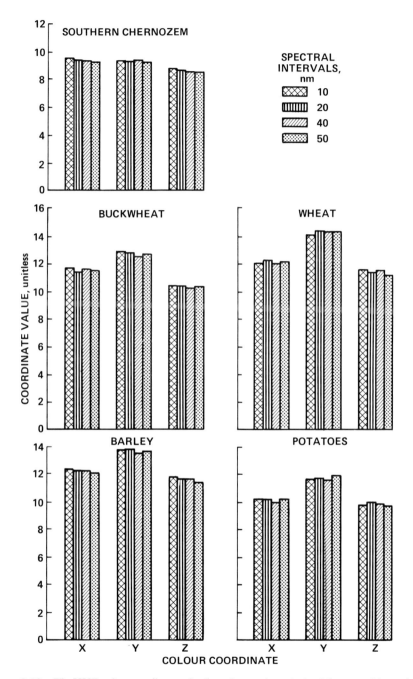

Figure 2.10 The XYZ *colour coordinates of soils and vegetation calculated for spectral intervals of 10, 20, 40 and 50 nm.*

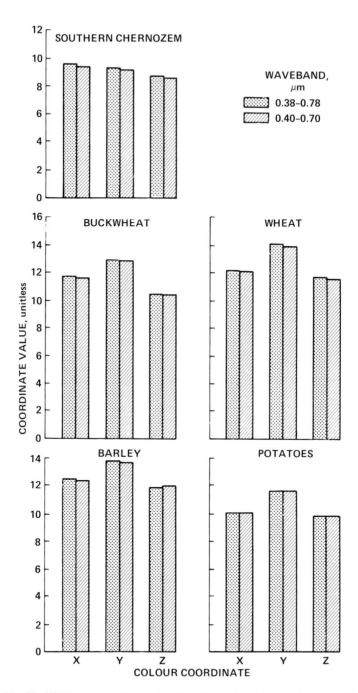

Figure 2.11 The XYZ *colour coordinates of soils and vegetation calculated for spectral intervals of* $0 \cdot 38 - 0 \cdot 78$ *and* $0 \cdot 40 - 0 \cdot 70$ μm.

2.8.3. Colour coordinate calculation using spectral intervals of different ranges

In colorimetry a colour is calculated from the SBC measured within the $0.38-0.78$ μm waveband. However, because the adding functions of the SCO are very low in the $0.38-0.40$ and $0.70-0.75$ μm spectral intervals, their contribution to the SBC calculation is small. Consequently, it is possible to narrow the spectral range to $0.40-0.70$ μm for calculating colour coordinates of soils and vegetation, without reducing the accuracy of their estimation. When this was done the error was found to be below 5 per cent (Figure 2.11).

2.9. Concluding comments on colour and its measurement

The transformation of remotely sensed measures of radiation into measures of colour is not a significant part of the remote sensing literature in the English language. However, it does have great potential for the study of colourful objects, like soils and vegetation, as is demonstrated throughout this book.

PART II
Spectral reflectance of soils

3
The theory of soil reflectance

The spectral reflectance of soils has been a research focus for a long time in the USSR (Obukhov and Orlov, 1964; Zyrin and Kuliev, 1967; Karmanov, 1974; Tolchelnikov, 1974; Orlov and Proshina, 1975; Orlov et al., 1976, 1978; Fedchenko and Kondratyev, 1981; Kondratyev and Fedchenko, 1980b, 1981a). In general, these studies have been empirical and little work has been done on the theoretical basis of radiation reflection from soils. Consequently, an understanding of the causes of reflection from soils has been inhibited in the USSR.

Recently three major approaches or approximations have been used in the theoretical modelling of the reflected radiation field from soils. First, the soil has been presented schematically in terms of either its coefficients of reflection for a flat interface by the Fresnel formulae or a geometric presentation of its heterogeneities by comparison with an idealised reflecting surface in the form of sets of cylinders and spheres of equal radii (discussed in Section 3.1). The second, and more realistic, approach uses the theory of radiation diffraction on a rough surface. Usually the single scattering theory is employed for this purpose. When the characteristic size of the roughness is much less than that of the wavelength of the incident radiation, the small-perturbations technique is used. Conversely, when the characteristic size of the roughness is much larger than the wavelength of the incident radiation, the tangent-plane technique is used (discussed in Section 3.2). An acoustic approach has also been tried and was found to be valuable in describing the interaction of light with a soil even though it represents the wave field in only one dimension (electromagnetic waves are three-dimensional in character), and ignores polarisation effects (Brekhovskikh, 1952a; Isakovich, 1952). The third and most realistic approximation is the multiple-scattering theory and the special case in this context is the theory of radiation transfer (discussed in Section 3.1). Results of the application of this theory for complicated randomly heterogeneous media such as soils and vegetation are discussed in Chapter 6.

All three approaches have their limitations and, as a result, inadequately represent the laws by which the reflected radiation field is formed. With the first approach the limited value of the Fresnel formulae is often apparent, and so, at least in the USSR, its use has been restricted to the contrast between smooth surfaces such as calm water and rough surfaces such as ploughed soil. Theoretical limitations to the second approach are also evident. The soil has not only an irregular surface structure, but also consists of optically heterogeneous particles of varying transparency, size, shape and composition (Brady, 1984). The third approach has many problems associated with the complicated composition of soils and vegetation and the need for complex mathematical description. Quantifying the correlation functions of heterogeneous distributions of the dielectric properties, radiation scatterers and surface roughness is difficult, as is the solution of the resulting integral equations. However, all three approaches do offer a description of the interaction of radiation and soil. Each is discussed in the next two sections.

3.1. A description of radiation–soil interaction using the laws of geometric optics

Soil is typically composed of various sizes of aggregates and illuminated by either global (direct plus scattered) or scattered solar radiation. In the theoretical calculation of soil reflection, one of the most important parameters to account for is the proportion of illuminated to shaded soil surface. In describing the laws of the formation of the reflected radiation field it is necessary to consider the diffraction effects of the interaction between the incident radiation field and a randomly heterogeneous soil cover. The theory of such an interaction has been developed only for extreme cases (Rytov et al., 1978). As a first approximation, however, geometric optics (Sharonov, 1958) may be used. To achieve this, the soil surface may be represented as one of three states—a smooth surface (e.g. uncultivated soil), a ridge and furrow surface (e.g. ploughed soil) or a surface of spheroid aggregates (e.g. harrowed soil).

3.1.1. The reflection of radiation by uncultivated soil

The reflection of light from an ideal smooth surface is specular and can be described by the Fresnel formulae (Kizel, 1973). Under the illumination of natural, unpolarised sunlight, the reflectance (brightness coefficient, r) of such a surface can be determined from (Born and Wolf, 1973)

$$r = \frac{1}{2}\left(\frac{\tan^2(\theta_i - \theta_t)}{\tan^2(\theta_i + \theta_t)} + \frac{\sin^2(\theta_i - \theta_t)}{\sin^2(\theta_i + \theta_t)}\right) \tag{3.1}$$

where θ_i and θ_t are the solar zenith angle and refraction angle, respectively, and the angle of reflection θ_r equals θ_i. Thus, for dense, uncultivated soils the angular distribution of reflected direct solar radiation $I(\theta)$ is characterised by discrete numbers (Figure 3.1). At view zenith angles equal to θ_i, reflectance is determined by the r parameter; at all others it is zero. The field of reflected radiation from the

The theory of soil reflectance

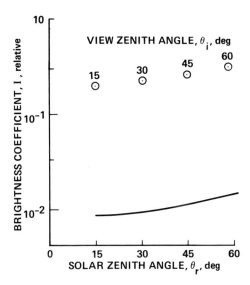

Figure 3.1 Angular distribution of solar radiation reflected from an ideal smooth surface for different view angles (points) and for scattered radiation (solid curve).

same surface, illuminated by scattered radiation, can be obtained by integrating Equation 3.1 over all incidence angles.

3.1.2. The reflection of radiation by ploughed soil

In Soviet remote sensing research the expression for the mean brightness of reflected radiation (Kulebakin, 1926) is often used to describe the reflection of radiation from a ploughed soil. In accordance with this, it is assumed that the illuminated surfaces reflect radiation isotropically, in keeping with Lambert's law, simply because it is difficult to describe non-isotropic reflection. The mean brightness $\bar{I}$ of a ploughed soil may be represented as

$$\bar{I} = \frac{c_l}{2 \sin \tilde{\gamma}_{max} \sin \theta} \{\tfrac{1}{4} \sin(\theta_i + \theta)[\cos 2\beta - \cos 2\theta_i]$$
$$+ \tfrac{1}{2} \cos(\theta_i + \theta)[2(\theta_i + \beta) + \sin 2\theta_i + \sin 2\beta] - (\theta_i + \beta)\cos \theta_i \cos \theta\} \quad (3.2)$$

where

$$c_l = r \frac{L_i}{\pi} l \quad (3.3)$$

l being the length of furrow within the instrument's field of view and L_i being the incident radiation flux. Other parameters are given in Figure 3.2, where β is the

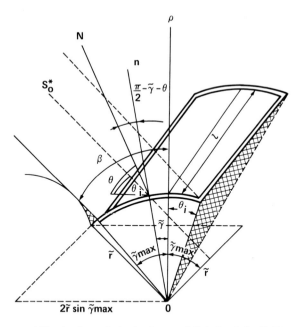

Figure 3.2   Geometry of illumination and viewing for a modelled ploughed soil. Furrows are modelled as cylindrical surfaces with radius $\tilde{r}$. Shaded parts of soil furrows are hatched.

angle characterising an unshaded side surface of the cylinder taken to represent a furrow, $\tilde{\gamma}_{max}$ is the maximum angle from the unshaded part of the furrow to nadir, S_0^* is the radiation source (Sun), N is the point of viewing, n is the normal to the surface with a radius of $\tilde{r}$ centred at point 0 and ρ is the nadir point of viewing. It follows from Equations 3.2 and 3.3 that the brightness coefficient is independent of the radius of the surface curvature.

Figure 3.3 illustrates the angular distributions of brightness coefficients for $\tilde{\gamma}_{max} = 60°$, $r = 1$, $l = 1$ and $L_i = 1$ for three solar zenith angles ($\theta_i = 15°$, $30°$ and $60°$). This shows how the brightness of a ploughed soil is positively related to the view zenith angle (Figure 1.1), a consequence of the soil surface curvature. This increase is due to the assumption of Lambertian reflection. It would not be observed at every azimuth angle in natural conditions. Figure 3.3 also shows that scattered radiation brightness is also positively related to the view zenith angle. Finally, from a comparison of Figures 3.3 and 3.1 it can be inferred that a stronger angular dependence of reflection is displayed by the ploughed than the uncultivated soil.

3.1.3. The reflection of radiation by harrowed soil

The reflection of light from a harrowed soil can be described with reference to the results of Kastrov (1955), where the soil aggregates were represented as equally

The theory of soil reflectance

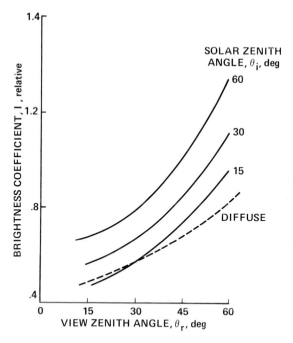

Figure 3.3 Angular distributions of solar radiation reflected from ploughed soil (ridge and furrow surface), for three solar zenith angles (15°, 30° and 60°) and for (diffuse) radiation (broken line).

sized spheres. The mean brightness under these conditions can be expressed as

$$\bar{I} = \frac{IL_i}{\sqrt{3}} \left(\cos\theta_i \int_0^{\alpha_1} \tilde{\Phi}(\theta') \frac{1+\sin\tilde{\phi}}{2} \cos\theta' \sin\theta' \, d\theta' \right.$$

$$+ \sin\theta_i \int_0^{\alpha_1} \tilde{\Psi}(\theta') \frac{1+\sin\tilde{\phi}}{2} \sin\theta' \, d\theta' - \cos\theta_i \tan\theta \int_0^{\alpha_2} \tilde{\Phi}'(\theta') \frac{1+\sin\tilde{\phi}}{2}$$

$$\left. \times \sin^2\theta' \, d\theta' - \sin\theta_i \tan\theta \int_0^{\alpha_2} \tilde{\Psi}'(\theta') \frac{1+\sin\tilde{\phi}}{2} \frac{\sin^3\theta'}{\cos\theta'} \, d\theta' \right) \quad (3.4)$$

where $\frac{1}{2}(1+\sin\tilde{\phi})$ is the proportion of the sphere's diameter that is illuminated, the angle $\tilde{\phi}$ characterizes the illuminated part of the spherical surface; $\tilde{\Phi}$, $\tilde{\Phi}'$, $\tilde{\Psi}$ and $\tilde{\Psi}'$ are functions related to the illuminated parts of the spherical triangles, and α_1 and α_2 are related to respective arcs of spherical surfaces illuminated by direct solar radiation. Equation 3.4 takes into account the number of spherical surfaces per unit area, in addition to the dependence of reflection on the inverse cosine between the view direction and the normal to a surface expressed as $(1 - \tan\theta \tan\theta' \cos\phi')$, where θ' and ϕ' are the polar angle and longitude, respectively. The functions $\tilde{\Phi}_z(\theta)$ and $\tilde{\Psi}_z(\theta)$ characterise the illuminated parts of spherical triangles and can be determined from Equation 3.5:

$$\text{with } 0 \leq \theta \leq \alpha'_1 \quad \tilde{\Phi}_z(\theta) = \pi, \quad \tilde{\Psi}_z(\theta) = 0$$
$$\text{with } \alpha'_1 \leq \theta \leq \alpha'_2 \quad \tilde{\Phi}_z(\theta) = \tilde{\phi}_1(\theta) = \arccos(\cot \theta, \cot \theta_i), \quad \tilde{\Psi}_z(\theta) = \sin \tilde{\phi}_1(\theta)$$
$$\text{with } \alpha'_2 \leq \theta \leq \alpha'_3 \quad \tilde{\Phi}_z(\theta) = \tilde{\phi}_1(\theta) - \tilde{\phi}_2(\theta) = \arccos(-\cot \theta \cot \theta_i)$$
$$- \arccos\left(\frac{\sin \varepsilon + \sin \theta_i \cos \theta}{\sin \theta \cos \theta_i}\right), \quad \tilde{\Psi}_z(\theta) = \sin \tilde{\phi}_1(\theta) - \sin \tilde{\phi}_2(\theta) \quad (3.5)$$

Here α_1, α_2 and α_3 are determined by the arcs of illuminated parts of spherical surfaces and ε is the position of the Sun's vertical plane with respect to the arc of a long circle. The functions $\tilde{\Phi}'(\theta)$ and $\tilde{\Psi}'(\theta)$ differ from $\tilde{\Phi}_z(\theta)$ and $\tilde{\Psi}_z(\theta)$ by the quantity $\tan \theta' \cos \phi'$.

It follows from Equation 3.4 that the brightness of the harrowed soil, composed of homogeneous spherical particles, is independent of the particle radius $(\tilde{r})$, but is determined by the number of particles per unit area, which is equal to $\frac{1}{2}\tilde{r}^2\sqrt{3}$. An exponential relationship between aggregate size and reflectance is sometimes observed, depending on the characteristic radius of the particles (Section 4.5).

Curves of the angular change in brightness coefficient derived from Equation 3.5 for $r = 1$ and $L_i = 1$ at three solar zenith angles ($\theta_i = 15°$, $30°$ and $60°$) are shown in Figure 3.4 for a harrowed soil. As with Figures 3.1 and 3.3, a positive relationship between brightness coefficient and view zenith angle is apparent. However, it is less than for the ploughed soil (Figure 3.3). Furthermore, when using scattered radiation, the recorded brightness coefficient was essentially independent of the view zenith angle.

All of the above calculations show how the laws of geometric optics may be used to study the process of radiation interaction with soils if certain environmentally unrealistic assumptions are made (e.g. isotropic reflection from illu-

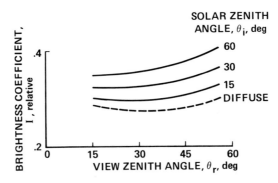

Figure 3.4 Angular distributions of solar radiation reflected from harrowed soil (spheroid aggregates) for three solar zenith angles (15°, 30° and 60°) and for diffuse radiation (broken line).

minated portions of a ploughed soil). However, the identified regularities of light reflection from soil concur with the calculations obtained from the theory of radiation transfer in Section 1.2 (compare Figures 3.1, 3.3, 3.4 and 1.3).

3.2. The theory of radiation reflection from rough surfaces (tangent-plane technique)

So far the theoretical basis for the reflection of light from a soil has been considered without taking into account the effects of coherence, spatial heterogeneity of the radiation fields and the proportions of illuminated and shaded soil. In this section the wave field is considered and the wave equation is solved with respective boundary conditions. Consequently, wave diffraction and interference, not considered in geometric optics, are discussed.

The theory of radiation transfer (Ross, 1975) is a special case in which the roughness autocorrelation radius (the distance at which any two points of the medium affect each other) is much shorter than the characteristic interval at which the radiation field is averaged (see Part 3). The averaging scale and dispersion of the distribution of the rough medium's heterogeneous dielectric (scattering and absorbing) properties must be much less than this interval. For the theory of radiation transfer the average field must vary at distances larger than the given scale of averaging. The medium must consequently lack a substantial attenuation of the field. A very rough medium is characterised by multiple scattering and resultant substantial attenuation of the field. Therefore, the theory of radiation transfer can only be used for rough and fairly rough media.

The theory discussed below applies to any medium. It allows for the statistical character of the scattering heterogeneities as well as shading and other effects not considered by the radiation transfer theory.

3.2.1. General solution

The equation for a rough surface can be expressed as $z = \zeta(x, y)$, where ζ is the differentiation function of the coordinates. This may either be a determined function such as a sinusoid (Brekhovskikh, 1952a,b) or a statistically uneven function (Antokolsky, 1948; Isakovich, 1952). To illustrate this, a simplification of the problem of reflection of a plane monochromatic wave from this surface is discussed. The wave has the potential Φ_0 and can be characterised by the vector $\mathbf{k}_0(k_x^0, 0, k_z^0)$, which defines the coordinate system, and the field of the diffracted wave is searched for at point $p(x, y, z)$. In the simplification of this problem the diffraction of a plane monochromatic wave can be considered first. This is similar to problems found in acoustics, and the solutions can be easily generalised to a three-dimensional electromagnetic wave. Here the time factor may be omitted as reference is made to a monochromatic wave of a definite frequency, ω.

The potential of the scattered wave at any point p can be determined from

Green's formula:

$$\Phi(p) = \frac{1}{4\pi} \int_s \int \left[\Phi \frac{\partial}{\partial n} \left(\frac{e^{ikR}}{R} \right) - \frac{e^{ikR}}{R} \frac{\partial \Phi}{\partial n} \right] ds \qquad (3.6)$$

where the integration is taken over the entire uneven surface; R (not to be confused with radiance here) is the distance between the point p and a point on the surface with a radius vector, $\mathbf{v}$. The differentiation is made along the normal, $\mathbf{n}$, to the surface, the position of which is characterised by direction cosines $\bar{\alpha}, \bar{\beta}, \bar{\gamma}$: $\mathbf{n} = \mathbf{n}(\bar{\alpha}, \bar{\beta}, \bar{\gamma})$.

The potential of the field at a random point on the surface is

$$\Phi(p) = e^{i k_0 v} + e^{i k v} V(\xi) \qquad (3.7)$$

where $V(\xi)$ is the coefficient of reflection of a plane wave depending on the angle ξ between $\mathbf{k}_0$ and $\mathbf{n}$.

By definition, on a scattering surface

$$\frac{\partial \Phi_0}{\partial n} = \bar{\alpha} \frac{\partial \Phi_0}{\partial x} + \bar{\beta} \frac{\partial \Phi_0}{\partial y} + \bar{\gamma} \frac{\partial \Phi_0}{\partial z} = i n k_0 \Phi_0 \qquad (3.8)$$

and in accordance with the law of reflection

$$\frac{\partial \Phi_{0\tau p}}{\partial n} = i n k_{0\tau p} \Phi_{0\tau p} = - i n k_0 V(\xi) \Phi_0 \qquad (3.9)$$

where $0\tau p$ means from p, thus the derivative of the total field at the boundary is

$$\frac{\partial \Phi}{\partial n} = \frac{\partial \Phi_0}{\partial n} + \frac{\partial \Phi_{0\tau p}}{\partial n} = i n k_0 [1 - V(\xi)] \Phi_0 \qquad (3.10)$$

An expansion of this to a spherical wave e^{ikR}/R was used by Brekhovskikh (1952a). Isakovich (1952) provided a solution for the Fraunhofer zone in which the quantity $1/R$ can be considered to be constant and equal to $1/R_0$, where R_0 is the distance to the point of observation from the coordinate origin, and the phase of the Green's function is presented as $kR = kR_0 - \mathbf{xv}$, where $\mathbf{x}$ is the vector directed to the point of observation equal to k by module. In both cases the solution to Equation 3.7 can be written in terms of the double integral

$$\Phi(p) \approx \frac{1}{4\pi} \iint \{ V(k - x)\mathbf{n} + (k + x)\mathbf{n} \} e^{i(k-x)v} ds \qquad (3.11)$$

In Brekhovskikh's (1952a) work Equation 3.11 is multiplied by expansion in plane waves. In Isakovich's (1952) work Equation 3.11 is multiplied by the quantity $i \exp(i k R_0)/R_0$. Elementary Huggens waves were determined by Antokolsky (1948) in the direction of specular reflection which accounts for the η coefficient characterising shading, where η is zero for shaded parts and unity for unshaded parts. In general this shading can be accounted for by multiplying Equation 3.11 by η.

At $V = 0$, only the second term in the subintegral expression remains and the potential $\Phi(p) \equiv 0$ (no reflection) depends on edge-effects and diffraction at surface boundaries. For a randomly rough surface $V = V(\xi)$ and an expansion of $V(\xi)$ in Fourier series may be used (Brekhovskikh, 1952a,b). Omitting the unimportant $i \exp(ikR_0)$ phase coefficient and taking $\mathbf{q} = \mathbf{k} - \mathbf{x}$,

$$\Phi(p) = \frac{1}{4\pi R_0} \mathbf{q} \iint n e^{i\mathbf{q}\mathbf{v}} \, ds \qquad (3.12)$$

Because $\mathbf{n}$ is a unity vector in the direction of the $d\mathbf{t}/ds$ vector, where $\mathbf{t} = d\mathbf{v}/ds$ is a unity tangent vector to the surface $z = \zeta(x, y)$ (Figure 3.5 shows an LL section of this surface with the plane y constant) and knowing that

$$\frac{\bar{\alpha}}{\bar{\gamma}} = -\frac{\partial \zeta'}{\partial x} = \zeta'_x \qquad \frac{\bar{\beta}}{\bar{\gamma}} = \frac{\partial \zeta}{\partial y} = -\zeta'_y$$

and

$$\mathbf{n} = \{\bar{\alpha}, \bar{\beta}, \bar{\gamma}\} = \{-\bar{\gamma}\zeta'_x; -\bar{\gamma}\zeta'_y; \bar{\gamma}\}$$

we obtain

$$\sqrt{\bar{\alpha}^2 + \bar{\beta}^2 + \bar{\gamma}^2} = \bar{\gamma}\sqrt{1 + \zeta'^2_x + \zeta'^2_y}$$
$$ds = dx \, dy \sqrt{1 + \zeta'^2_x + \zeta'^2_y}$$
$$\mathbf{n} = (-\mathbf{i}_x \zeta'_x - \mathbf{i}_y \zeta'_y + \mathbf{i}_z)/\sqrt{1 + \zeta'^2_x + \zeta'^2_y}$$

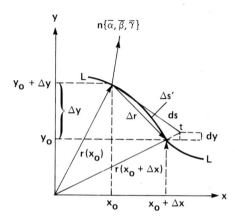

Figure 3.5 Section LL by the y = constant plane (refer to Equation 3.12).

where $\mathbf{i}_x$, $\mathbf{i}_y$ and $\mathbf{i}_z$ are unit vectors along the coordinate axes. This gives

$$\iint n e^{iqv}\, ds = \iint (-\mathbf{i}_x \zeta'_x - \mathbf{i}_y \zeta'_y + \mathbf{i}_z) e^{iqv}\, dx\, dy \tag{3.13}$$

Since

$$\frac{\partial}{\partial x} e^{iqv} = (iq_x + iq_z \zeta'_x) e^{iqv} \tag{3.14}$$

$$\frac{\partial}{\partial y} e^{iqv} = (iq_y + iq_z \zeta'_y) e^{iqv} \tag{3.15}$$

upon excluding ζ'_x and ζ'_y from Equation 3.13, 3.14 and 3.15,

$$\iint n e^{iqv}\, ds = \iint \left(\mathbf{i}_x \frac{q_x}{q_z} + \mathbf{i}_y \frac{q_y}{q_z} + \mathbf{i}_z - \frac{\mathbf{i}_x}{iq_z}\frac{\partial}{\partial x} - \frac{\mathbf{i}_y}{iq_z}\frac{\partial}{\partial y} \right) e^{iqv}\, dx\, dy \tag{3.16}$$

If the edge-effects arising from the last two components of Equation 3.16 are neglected, then

$$\iint n e^{iqv}\, ds = \frac{\mathbf{q}}{q_z} \iint e^{iqv}\, dx\, dy \tag{3.17}$$

and the scattered field potential may be expressed as

$$\Phi(p) = \frac{1}{4\pi R_0} \frac{q^2}{q_z} \iint e^{iqv}\, dx\, dy \tag{3.18}$$

Like Asmus et al. (1980), shading can be accounted for by multiplying Equation 3.18 by η and taking the three-dimensional function of the distribution of roughness and derivatives ζ'_x and ζ'_y.

3.2.2. Special solution of the problem for unlimited sinusoids

Brekhovkikh (1952b) shows the solution for this case is

$$\Phi(p) = \sum_{m,n=-\infty}^{+\infty} B_{mn}(k_x^m, k_y^n, k_z^{mn}) \exp[i(k_x^m x + k_y^n y + k_z^{mn} z)] \tag{3.19}$$

where a set (spectrum) of wavevectors (numbers) in a scattered wave along the x-axis (the xz-plane coincides with the plane of incidence) is equal to the sum of wavenumbers of the k_0 incident wave, and a set of periods of sinusoids P: along x

$$k_{x'}^m = k_x^0 + mP$$

along y

$$k_y^n = nq \quad (q \text{ is the } y \text{ period of sinusoids})$$

along z

$$k_z^{mn} = \sqrt{k^2 - (k_x^m)^2 - (k_y^m)^2}$$

The $B_{mn}(k_x, k_y, k_z)$ coefficients are estimated from the expansion

$$\frac{1}{2k_z}[(k_z - k_z^0) - \zeta_x'(k_x - k_x^0) - \zeta_y'k_y]V(\zeta)\exp[i(k_z^0 - k_z)\zeta]$$

$$= \sum_{m,n=-\infty}^{+\infty} B_{mn}(k_x, k_y, k_z)\exp[i(mPx + nqy)] \quad (3.20)$$

If account is taken of surface roughness (V = constant; $V = 1$ for a rough surface, $V = -1$ for a smooth surface), then

$$\Phi(p) = V \sum_{m,n=-\infty}^{+\infty} \frac{k^2 - k_z^{mn}k_z^0 - k_x^m k_x^0}{k_z^{mn}(k_z^{mn} - k_z^0)} B_{mn}^0(k_z^{mn})\exp[i(k_x^m x + k_y^n y + k_z^{mn} z)] \quad (3.21)$$

where $B_{mn}^0(k_z)$ are the coefficients of the expansion in the Fourier series

$$\exp[i(k_z^0 - k_z)\zeta(x,y)] = \sum_{m,n=-\infty}^{+\infty} B_{mn}^0(k_z)\exp[i(mPx + nqy)] \quad (3.22)$$

B_{mn} being expressed in terms of B_{mn}^0 by

$$B_{mn}(k_x, k_y, k_z) = \{V/[k_z(k_z - k_z^0)]\}(k^2 - k_z k^0 - k_x k_x^0)B_{mn}^0(k_z) \quad (3.23)$$

This shows that the solution is determined by an infinite number of 'lobes' (spectral harmonics) which include sets of wavenumbers of the incident wave and sets of periods of sinusoids.

3.2.3. Special solution for the problem for limited sinusoids

If a sinusoid is limited, for instance, along the x-axis with the illumination function $F(x)$ equal to unity on the sinusoid and zero outside it, then instead of Equation 3.19 (with the sinusoid assumed to be unlimited along the y-axis),

$$\Phi(p) = \sum_{m=-\infty}^{+\infty} \int_{-\infty}^{+\infty} B_m(k_x, k_z) T(k_x^0 - k_x + mP)\exp[i(k_x x + k_z z)] \, dk_x \quad (3.24)$$

applies with the sinusoid unlimited along the x-axis, instead of the T function there will be a δ function of the $(k_x^0 - k_x + mP)$ argument, which will lead to a definite correlation between wavenumbers k_z and periods of P sinusoids along the x-axis. Here, in the case of a limited site size, there will be waves with random k_x (i.e. waves propagating in every direction). However, for each of m amplitudes of these waves within the range $k_x = k_x^0 + mP \pm \Delta\chi$ (where $\Delta\chi$ is the interval of variation of the derivative in the expansion 3.25) they will be markedly different from zero:

$$\left. \begin{array}{l} F(x) = \displaystyle\int_{-\infty}^{+\infty} e^{-1\chi_x} T(\chi) \, dx \\[2ex] T(\chi) = \dfrac{1}{2\pi} \displaystyle\int_{-\infty}^{+\infty} e^{i\chi_x} F(x) \, dx \end{array} \right\} \quad (3.25)$$

Because $\Delta\chi \ll P$, the scattered waves are grouped in narrow angular intervals near the disection prescribed by

$$k_x = k_x^0 + mP \qquad m = 0, \pm 1, \pm 2, \ldots \qquad (3.26)$$

In the Fraunhofer zone the solution can be written as

$$\Phi(p) = B_m(\tilde{\theta}) \sin \tilde{\alpha} T(k_x^0 + mP - k \cos \tilde{\theta}) \sqrt{\dfrac{\pi k}{R}} \exp\left(-\dfrac{i\pi}{4} + ikR\right) \quad (3.27)$$

where $\tilde{\theta}$ is the angle between the direction to the observation point and the x-axis. Here the diffraction field is a cylindrically scattering wave with the directional characteristic in the shape of lobes, the maxima of which lie in directions $\tilde{\theta} = \tilde{\theta}_m$ determined from

$$k \cos \tilde{\theta}_m = k_x^0 + mP \qquad m = 0, \pm 1, \pm 2, \ldots \qquad (3.28)$$

The angular width of the lobes can be described by

$$\tilde{\theta}_m - \tilde{\theta}_m' \approx \lambda / \tilde{D} \sin \tilde{\theta}_m \qquad (3.29)$$

where $\tilde{D}$ is the interval of variations in the x variable; at a given $F(x)$ the wave amplitude in the centre of the lobes is

$$A_m = \left(\dfrac{\tilde{D}}{2\pi}\right) \sqrt{\dfrac{\pi k}{R}} B_m(\tilde{\theta}_m) \sin \tilde{\theta}_m \qquad (3.30)$$

where the $B_m(\tilde{\theta}_m)$ are estimated from Equation 3.23.

Thus, in this case the solution is presented as a set of discrete lobes of a certain width and amplitude. The angular dependence of scattered waves can be determined from

$$f(\tilde{\theta}_m) = |B_m(\tilde{\theta}_m)| \sin \tilde{\theta}_m \qquad (3.31)$$

If $|V| = 1$, then, using the angle of sliding incident wave and the expansion

$$\exp[-iR(\sin \tilde{\theta}_m + \sin \tilde{\theta}_0)\zeta(x)] = \sum_{m=-\infty}^{+\infty} B_m^0(\tilde{\theta}_m) e^{imPx} \qquad (3.32)$$

where $\zeta(x) = \zeta_0 \cos Px$,

$$f(\tilde{\theta}_m) = \frac{1 - \cos(\tilde{\theta}_m - \tilde{\theta}_0)}{\sin \tilde{\theta}_m + \sin \tilde{\theta}_0} J_m\left(\frac{2\pi}{\lambda} \zeta_0(\sin \tilde{\theta}_m + \sin \tilde{\theta}_0)\right) \qquad (3.33)$$

is obtained. Brekhovskikh (1952b) provides graphs illustrating the dependence of diffracted waves on solar zenith angle for a variety of soil roughnesses and wavelengths. Figures 3.6–3.8 show the field of diffracted radiation for three solar zenith (incidence) angles for the uncultivated, ploughed and harrowed soils outlined in Sections 3.1.1–3.1.3.

For the smooth soil surface the law of reflection was close to the specular one, especially at large solar zenith angles. Decreasing the solar zenith angle resulted in a broadening of the angular anisotropy of diffracted radiation with the respective lobes (discrete modes of reflection) located symmetrically around the major specular peak (Figure 3.6).

With the ploughed soil the law of reflection differed markedly from the specular one, but it was still distinctly non-Lambertian (Figure 3.7). Furthermore, the angular distribution of the lobes was only weakly dependent on the angle of incidence. When actually viewing such a surface at different angles, a relatively weak reflection (in the directions between the lobes) and a relatively strong reflection (in the region of the lobes' maxima) were apparent. These modes of reflection are more regular in both amplitude and angular distribution for the harrowed soil, this medium giving a smoother angular anisotropy of reflection that approaches Lambertian reflectance (Figure 3.8).

Comparing Figure 3.6–3.8 with Figures 3.1, 3.3 and 3.4 reveals that a rather unrealistic angular distribution of diffracted radiation is provided by using geometric optics. For instance, the smooth soil surface in Figure 3.6 shows reflection at angles other than specular, and Figures 3.7 and 3.8 show non-Lambertian reflection, contrary to the assumption made in geometric optics. Consequently, the theory of radiation diffraction on a rough surface may be considered as a better approximation than geometric optics. However, it does not consider multiple scattering. A more general theory of the interaction of radiation with randomly heterogeneous media which does account for multiple scattering is given in Chapter 6.

Spectral reflectance of soils

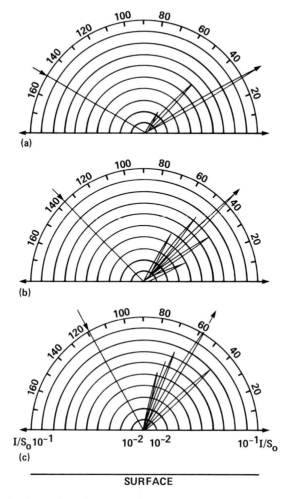

Figure 3.6 Angular distributions of solar radiation reflected from uncultivated soil (smooth surface) for three solar zenith angles: (a) 60°, (b) 45° and (c) 30°.

3.2.4. Special solution to the problem in the case of a statistically rough surface

In addition to the discrete patterns previously described, an irregular surface produces scattered background radiation at every view angle, obscuring the lobe-like reflection pattern shown in Figures 3.6–3.8. If the interface $z = \zeta(x, y)$ is of random character, it is possible to average the scattered radiation expression (Equation 3.18) and form a characteristic function

$$\overline{e^{\mathrm{i} q_z \zeta}} = \tilde{f}(q_z) = \int e^{\mathrm{i} q_z \zeta} \widetilde{W}(\zeta) \, \mathrm{d}\zeta$$

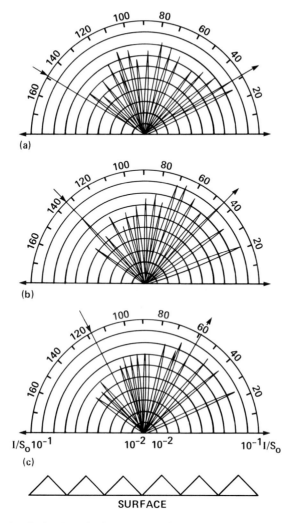

Figure 3.7 Angular distributions of solar radiation reflected from a ploughed soil (ridge and furrow surface) for three solar zenith angles: (a) 60°, (b) 45° and (c) 30°.

where $\widetilde{W}(\zeta)$ is the probability density of ζ. The function can be expressed, if necessary, in terms of statistical moment. Thus, with any one of the probability density, the characteristic function or the roughness function moment known, the average of the field scattered by a random rough surface, $\tilde{\Phi}$, can be calculated. The average field, $\tilde{\Phi}$, is the product of the $\tilde{f}(q_z)$ function by the potential of the field scattered by the same surface when not rough, the factor with the characteristic function differing from zero only within narrow angles close to the

Spectral reflectance of soils

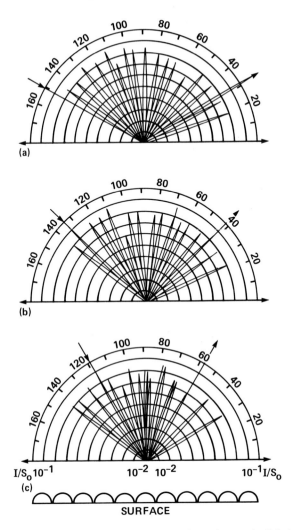

Figure 3.8 Angular distributions of solar radiation reflected from a harrowed soil (spheroid aggregates) for three solar zenith angles: (a) 60°, (b) 45° and (c) 30°.

angle of specular reflection. This arises because each of the integrals

$$\int \exp[(k_x - \chi_x)x]\, dx \quad \text{and} \quad \int \exp[(k_y - \chi_y)y]\, dy$$

approaches the δ function from $k_x - \chi_x$ and $k_y - \chi_y$, with these integrals approaching $\pm \infty$. In the specular direction

$$\bar{\Phi} = -(ks/2\pi R_0)\cos\theta \tilde{f}(-2k\cos\theta) \qquad (3.34)$$

The theory of soil reflectance

where s is the area of a scattering surface and $\tilde{f}(2k\cos\theta)$ is a result of surface roughness.

The derivatives ζ'_x and ζ'_y were excluded from Equations 3.17 and 3.18 as these, like the $z = \zeta(x, y)$, are random functions. However, in a general case these random derivatives must be accounted for, although the resulting expression is more cumbersome than before. In the case of specular reflection

$$\bar{\Phi} = \frac{ik(\bar{\gamma}_1 + \bar{\gamma}_2)}{4\pi} \iint \frac{e^{ik(R_1 + R_2)}}{R_1 R_2} \tilde{f}(\mathbf{R}_1, \mathbf{R}_2)\, dx\, dy \qquad (3.35)$$

where $\mathbf{R}_1 = \{\bar{\alpha}_1, \bar{\beta}_1, \bar{\gamma}_1\}$, $\mathbf{R}_2 = \{\bar{\alpha}_2, \bar{\beta}_2, \bar{\gamma}_2\}$ and the $\tilde{f}(\mathbf{R}_1, \mathbf{R}_2)$ function characterises the result of averaging scattered (Huggens) waves in $F(z, dz/dx, dz/dy)$. Therefore, at a given direction from the scattering site

$$\bar{\Phi} = \tilde{\Phi}\tilde{f}(2k\bar{\gamma}_0) \qquad (3.36)$$

which coincides with Equation 3.34 and $\bar{\gamma}_0$ corresponds to a geometrically reflected beam.

With the function $\tilde{F}$ assumed to be even with respect to dz/dy and dz/dx, an inverse Fourier transformation of the measured function $\tilde{f}(k)$ enables the function of the distribution of surface roughness and rough surface slopes to be expressed as

$$\tilde{F}_g(z) + \frac{\bar{\alpha}_1 + \bar{\alpha}_2}{\bar{\gamma}_1 + \bar{\gamma}_2}\bar{\nu}_\zeta(z)$$

where $\bar{\nu}_\zeta$ is an average of dz/dx at a fixed value of $z = \zeta$. As described in Section 3.2.1, shading can be accounted for with the η function which gives a shading-modulated function F_g with respect to dz/dy, but not to dz/dx.

The quantity $\tilde{f}$, which characterises reflection, can be calculated for a given $\tilde{F}_g(\zeta)$ by

$$\tilde{f}(k, \bar{\gamma}_0) = e^{-k^2\bar{\gamma}_0\beta_0/2}$$

where

$$\beta_0 = -\tilde{f}''(0) = \overline{z^2} + \frac{\bar{\alpha}_1 + \bar{\alpha}_2}{\bar{\gamma}_1 + \bar{\gamma}_2} \int_{-\infty}^{\infty} z^2 \overline{\nu_z(z)}\, dz$$

If the average potential of a random surface (Equation 3.34) is maximised at directions close to specular reflection, the average intensity or mean square of the field and the intensity fluctuation will also be large at non-specular angles.

The mean intensity of a 'cavity' with a unit width in the y-axis at an azimuth angle $\phi = 0$ is

$$\overline{\Phi\Phi^*} = \frac{1}{16\pi^2 R_0^2} \frac{q^4}{q_z^2} \iint e^{iq_x(x-x_1)} \overline{e^{iq_z(\zeta-\zeta_1)}} \, dx \, dx_1 \qquad (3.37)$$

where $\zeta = \zeta(x), \zeta_1 = \zeta(x_1)$, and the asterisk denotes complex conjunction. Substituting $\tau = x_1 - x$, assuming that the correlation radius τ is very small relative to the size of the surface ζ gives

$$\overline{\Phi\Phi^*} = \frac{s}{16\pi^2 R_0^2} \frac{q^4}{q_z^2} \int \overline{e^{iq_x\tau}} \overline{e^{iq_z(\zeta-\zeta_1)}} \, d\tau \qquad (3.38)$$

Here the double integral has been reduced to a one-dimensional one and s can be placed outside the integral (Kondratyev et al., 1986a). Here

$$\overline{e^{iq_z(\zeta-\zeta_1)}} = \iint e^{iq_z(\zeta-\zeta_1)} \widetilde{W}(\zeta,\zeta_1) \, d\zeta \, d\zeta_1 = \tilde{f}(q_z, -q_z) \qquad (3.39)$$

is the characteristic function, $\widetilde{W}(\zeta,\zeta_1)$ is the probability density distribution for a two-dimensional random quantity (ζ, ζ_1), and $\widetilde{W}$ and $\tilde{f}$ are functions of τ. The characteristic function of the two-dimensional quantity may also be expressed in terms of successive moments. Thereby, the mean intensity can be calculated with known moments of the (ζ, ζ_1) function, or $\tilde{f}(q_z, -q_z)$ characteristic function, or $\widetilde{W}(\zeta, \zeta_1)$ probability density.

From Equations 3.38 and 3.39,

$$\overline{\Phi\Phi^*} = \frac{s}{16\pi^2 R_0^2} \frac{q^4}{q_z^2} \int e^{-iq_x\tau} \tilde{f}(q_z, -q_z) \, d\tau \qquad (3.40)$$

At view zenith angles away from that of specular reflection, the intensity fluctuations (the difference $\overline{\Phi\Phi^*} - \bar{\Phi}\bar{\Phi}^*$ between the mean square of the field and the square of average field) coincides with mean intensity because the average potential at very non-specular reflection angles is almost zero. However, with large roughness amplitudes, when specular reflection is very small, this term may be neglected.

From Equations 3.36 and 3.40, and noting that $\tilde{f}^*(t) = \tilde{f}(-t)$,

$$\overline{\Phi\Phi^*} - \bar{\Phi}\bar{\Phi}^* = \frac{s}{16\pi^2 R_0^2} \frac{q^4}{q_z^2} \int e^{iq_x\tau}[\tilde{f}(q_z, -q_z) - \tilde{f}(q_z, 0)\tilde{f}(0, -q_z)] \, d\tau \qquad (3.41)$$

For the normal (Gaussian) two-dimensional distribution

$$\widetilde{W}(\zeta,\zeta_1) = \frac{1}{2\pi\tilde{\sigma}^2\sqrt{1-\tilde{r}^2}} \exp\left(-\frac{1}{2\tilde{\sigma}^2(1-\tilde{r}^2)}(\zeta^2 - 2\tilde{r}\zeta\zeta_1 + \zeta_1^2)\right)$$

where $\tilde{r} = \tilde{r}(\tau) = \overline{\zeta\zeta_1}/\bar{\zeta}^2$ is the autocorrelation coefficient for the quantity ζ and $\tilde{\sigma}^2$ is the dispersion

$$\overline{\Phi\Phi^*} = \frac{s}{16\pi^2 R_0^2} \frac{q^4}{q_z^2} \int \exp\{-iq_x\tau - q_z^2\tilde{\sigma}^2[1 - \tilde{r}(\tau)]\}\, d\tau$$

If the correlation coefficient can be represented as $\tilde{r}(\tau) = e^{-\tau^2/m^2}$ (usually used in supposition) then, since

$$\tilde{r}'(\tau) = \frac{2\tau}{m^2}\tilde{r}(\tau) \qquad \tilde{r}''(\tau) = -\frac{2}{m^2}\tilde{r}(\tau) + \frac{4\tau^2}{m^4}\tilde{r}(\tau)$$

$$\tilde{r}''(0) = -\frac{2}{m^2} \qquad \overline{\zeta'^2} = \frac{2\tilde{\sigma}^2}{m^2}$$

then

$$I = \overline{\Phi\Phi^*} = \frac{s}{16\pi^{3/2} R_0^2} \frac{q^4}{\sqrt{2q_z^3}} \frac{m}{\tilde{\sigma}} \exp\left(-\frac{1}{4}\frac{m^2}{\tilde{\sigma}^2}\frac{q_x^2}{q_z^2}\right) \qquad (3.42)$$

Figure 3.9 illustrates the calculated angular distributions of the intensity of radiation reflected from a rough surface, normalised to the solar constant, for different $m/\tilde{\sigma}$ ratios. These curves were derived from Equation 3.42, and their smooth character contrasts with the discrete narrow lobes in angular variation of reflected radiation from a sinusoidal surface. These curves also correspond more closely to reality than the idealised data portrayed in Figures 3.6–3.8.

From Figure 3.9 it is clear that the larger the ratio (therefore, the smoother the surface and the smaller the mean square of the $\overline{\zeta_x'^2}$ derivative) is, the closer the reflection comes to being specular. Larger ratios may occur when there is a decreasing autocorrelation (as the heterogeneities are far apart, producing a surface that is, on average, only slightly rough) or when there are many random surface slopes of different orientations that are smooth.

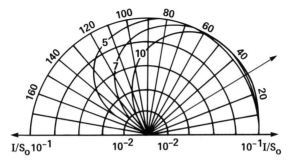

Figure 3.9 Angular distributions of solar radiation reflected from a rough surface for a 60° solar zenith angle at three values of the $m/\tilde{\sigma}$ ratio (5, 7 and 10).

3.3. Concluding comments on the theory of soil reflectance

This chapter has provided the current theory of soil reflectance as seen through the eyes of Soviet researchers. Such theory is the basis upon which much current work on the remote sensing of soils is being undertaken in the USSR. Chapters 4, 5 and 11 examine the reflectance properties of soils in the laboratory and field, and how theory and measurements can be combined to help us to understand soil reflectance.

4
Reflectance of soils in the laboratory and field

In general soil reflectance (SBC) is low but increases monotonically with wavelength through the visible and near-infrared portions of the electromagnetic spectrum (Obukhov and Orlov, 1964; Orlov 1966a, 1969; Orlov et al., 1966, 1976; Zyrin and Kuliev, 1967; Mikhailova, 1970; Karmanov, 1974; Tolchelnikov, 1974; Orlov and Proshina, 1975; Korzov and Ter-Markaryants, 1976; Sadovnikov and Orlov, 1978; Kondratyev and Fedchenko, 1980b, 1981a; Fedchenko and Kondratyev, 1981; Rachkulik and Sitnikova, 1981). In this chapter the spectral reflectance of soils and soil-forming rocks observed under laboratory and field conditions is discussed.

4.1. Spectral reflectance of topsoils

In most studies of soil reflectance only the topsoil, or upper horizon, has been considered, yet a wide range of spectral-reflectance curves has been obtained. This is, in part, due to natural variations, but some inconsistency has arisen from the use of a variety of experimental approaches and equipment. To allow objective comparison of soil reflectances, many soil samples were obtained from a region to the south of Moscow (Figure 4.1) and their SBCs were calculated in a standardised way. In general the soil samples were obtained from the upper horizon of newly ploughed arable fields in the Ukraine. These were air-dried, chopped and sieved (0·5 and 1·0 mm sieves) to remove the influence of soil moisture (Fedchenko and Kondratyev, 1981; Rachkulik and Sitnikova, 1981) and roughness (Kondratyev and Fedchenko, 1980c) on laboratory measurements of spectral reflectance.

Predictably, the SRCs differed markedly between the soil types, although each displayed the positive relationship between the SRC and wavelength noted in other studies (Figure 4.2, pp. 65–69). The nature of this relationship was variable between soils. A linear relationship was evident for dark soils (chernozems, brown soils, dark-grey forest soils, etc.) whereas an almost parabolic

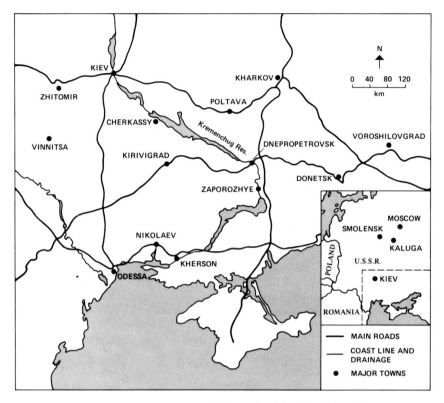

Figure 4.1   Locations in the USSR mentioned in this and later chapters.

relationship was apparent for light soils. In general the highest SRCs were observed for peat–podzol and turf–gley soils, and the lowest SRCs were observed for chernozem soils. This latter factor is mainly a function of the soil's humus content (Chapter 5).

4.2. Spectral reflectance of soils by horizon

Although remote sensing is primarily concerned with the surface soil, the spectral reflectance of soil from lower horizons can be of interest. For instance, many agricultural and engineering tasks produce much soil reorganisation, and soil from lower horizons is brought to the surface.

The depth-dependence of the soil SRC was investigated on samples taken from a chernozem soil in a silage pit near Kiev. As before, these samples were air-dried, chopped and sieved before measuring the SRC with an SF-18 spectrophotometer in the laboratory. Down to a depth of 60 cm the soil SRC varied little with distance from the surface or wavelength. At depths greater than 60 cm a sharp increase in SRC with depth at almost all wavelengths was apparent, as was a slight bend in the reflectance curve around the 0·50–0·60 μm, region (Figure 4.3). The highest SRC was observed from the soil-forming rock, a porous loess. Soils with different depths and properties will, of course, display different relationships.

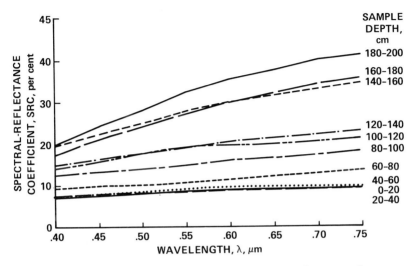

Figure 4.3 Spectral characteristics, by horizon, for a chernozem soil.

---

*Figure 4.2 (overleaf) Spectral characteristics of 21 peat–podzol soils and 25 chernozem soils from the Ukraine. Peat–podzol soils: (**1**) peat–gley medium-loamy soil (Chernigov); (**2**) peat–gley light-loamy soil (Chernigov); (**3**) peat–medium podzol light-loamy soil (Chernigov); (**4**) peat–podzol gley light-loamy soil (Chernigov); (**5**) peat–weak podzol light-loamy soil (Kiev); (**6**) peat–podzol medium-loamy soil (Kiev); (**7**) peat–podzol heavy-loamy soil (Kiev); (**8**) peat–podzol sandy loam (Kiev); (**9**) peat–podzol sandy soil (Zhitomir); (**10**) peat–medium podzol medium-loamy soil (Zhitomir); (**11**) peat–weak podzol sandy soil (Zhitomir); (**12**) peat–medium podzol sandy soil (Zhitomir); (**13**) peat–podzol light-loamy soil (Kiev); (**14**) peat–weak podzol sandy soil (Kiev); (**15**) peat–podzol light-loamy soil (Kiev); (**16**) peat–medium podzol sandy soil (Chernigov); (**17**) peat–podzol medium-loamy soil (Chernigov); (**18**) light-grey forest sandy loam (Vinnitsa); (**19**) grey forest sandy loam (Vinnitsa); (**20**) dark-grey forest sandy loam (Vinnitsa); (**21**) dark-grey podzolized medium-loamy soil (Khmelnitsk). Chernozem soils: (**1**) chernozem, grey forest medium-loamy soil (Khmelnitsk); (**2**) chernozem, grey forest sandy loam (Khmelnitsk); (**3**) chernozem, brown heavy-loamy soil (Crimea); (**4**) chernozem, dark-brown medium-loamy soil (Crimea); (**5**) chernozem, dark-brown sandy loam (Crimea); (**6**) chernozem, dark-brown heavy-loamy soil (Crimea); (**7**) chernozem, dark-brown heavy-loamy soil (Crimea); (**8**) chernozem, dark-brown heavy-loamy soil (Crimea); (**9**) chernozem, dark-brown heavy-loamy soil (Crimea); (**10**) chernozem, dark-brown heavy-loamy soil (Crimea); (**11**) chernozem meadow medium-loamy soil (Kiev); (**12**) usual chernozem light-loamy soil (Kirovograd); (**13**) south chernozem medium-loamy soil (Odessa); (**14**) south chernozem heavy-loamy soil (Zaporozhye); (**15**) south chernozem medium-loamy soil (Zaporozhye); (**16**) usual chernozem heavy-loamy soil (Kirovograd); (**17**) usual chernozem medium-loamy soil (Kirovograd); (**18**) usual chernozem heavy-loamy soil (Zaporozhye); (**19**) chernozem meadow light-loamy soil (Chernigov); (**20**) usual chernozem light-loamy soil (Zaporozhye); (**21**) thick chernozem light-loamy soil (Kirovograd); (**22**) thick chernozem medium-loamy soil (Cherkassy); (**23**) thick chernozem light-loamy soil (Kiev); (**24**) thick chernozem medium-loamy soil (Kiev); (**25**) thick chernozem heavy-loamy soil (Cherkassy).*

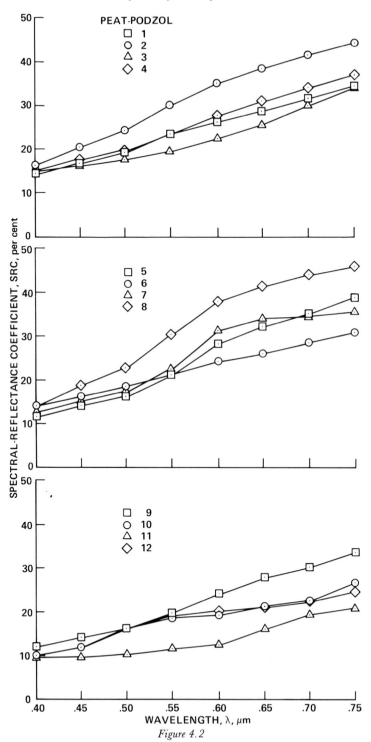

Figure 4.2

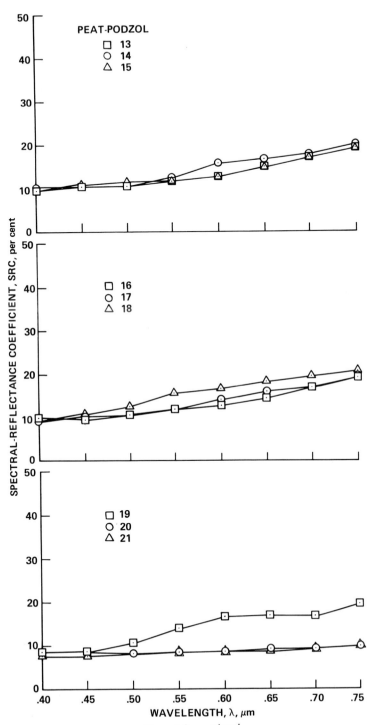

Figure 4.2 continued

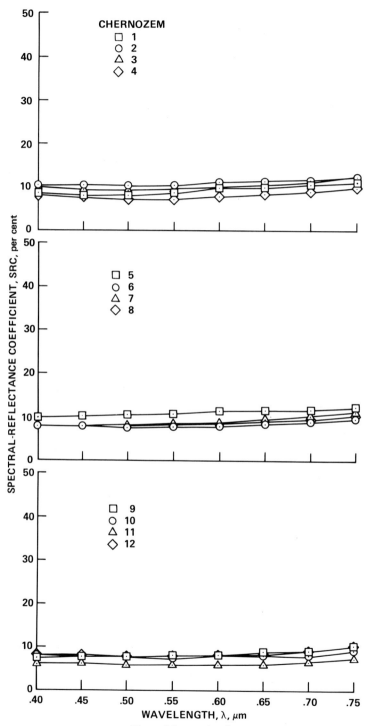

Figure 4.2 continued

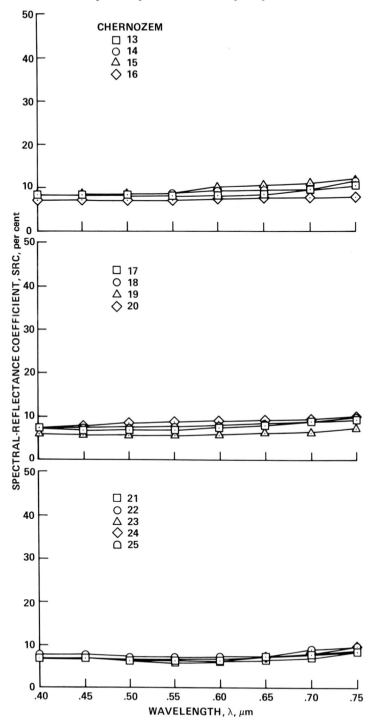

Figure 4.2 continued

4.3. Spectral reflectance of soil-forming rocks

To establish a quantitative relationship between the properties of soil and its colour, a knowledge of the spectral reflectance of the soil-forming rock is necessary. In the study areas the predominant rock type is loess, a typically loose, straw-coloured material.

The loess samples were obtained primarily from the central Ukraine (Kiev, Poltava, Kirivigrad, Cherkassy, Vinnitsa and Zhitomir), with some samples from hydrometeorological stations in other regions (Voroshilovgrad, Zaporozhye, Nikolaev and Kherson). These samples were subjected to the same analytical procedure used in Sections 4.1 and 4.2. Although the loess differ in terms of their mechanical composition and colour, the SRCs were relatively similar (Figure 4.4), the level of variation was no more than for a soil's upper or lower horizons. The level of variation was low, but the magnitudes of the loess SRCs were much higher than those observed from the upper horizon of the soils, as would be expected from Figure 4.3.

4.4. Spectral reflectance of soils from field measurements

Field investigations of soil spectral reflectance have often given results comparable with laboratory measurements of soil spectral brightness. However, the SBC measured in the laboratory is, in general, higher than that measured under field conditions because of a difference in environmental conditions and analytical procedures. Laboratory measurements of SBC are typically made using a spectrophotometer with an integrating sphere on processed (dried, sieved, etc.) soil samples. By contrast, field measurements of soils are made on unprocessed samples with radiometers with limited field-of-view. The magnitude of the SBC is therefore strongly dependent on the reflection indicatrix of the soil, and on such external factors as illumination conditions, moisture content and degree and characteristics of cultivation. Consequently, it is not possible to apply conclusions

---

Figure 4.4 Spectral characteristics of 22 loess samples from 10 regions of the USSR. (1) Yellow-straw-coloured, with greyish-brown patches (Kiev); (2) light-straw-coloured, loose (Vinnitsa); (3) whitish-yellow, dense (Zhitomir); (4) straw-coloured grey, sometimes with carbonate mould (Vinnitsa); (5) straw-coloured, with sand (Kiev); (6) yellow-brown, with light patches (Cherkassy); (7) greyish brown straw-coloured, dense (Kirivigrad); (8) straw-coloured, with yellow tint (Cherkassy), (9) light-brown, sometimes with 'white-eye' soil (Chernigov); (10) light-straw-coloured, with yellow-brown patches (Kiev); (11) dark-straw coloured, dense (Chernigov); (12) yellow, dense, with sand (Zhitomir); (13) yellow-straw-coloured, loose (Kherson); (14) straw-coloured, sometimes with carbonate mould (Poltava); (15) straw-coloured, slightly dense, sometimes with 'white-eye' soil (Poltava); (16) straw-coloured, very dense (Voroshilovgrad); (17) straw-coloured, very dense (Voroshilovgrad); (18) dark, with reddish tint (Poltava); (19) light-brown, sometimes with 'white-eye' soil (Cherkassy); (20) yellow, dense, with sand (Nikolaev); (21) straw-coloured, porous (Kherson); (22) light-brownish straw-coloured, porous (Zaporozhye).

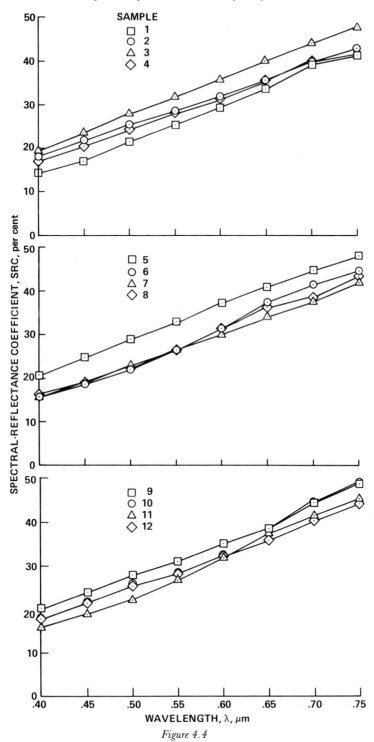

Figure 4.4

Spectral reflectance of soils

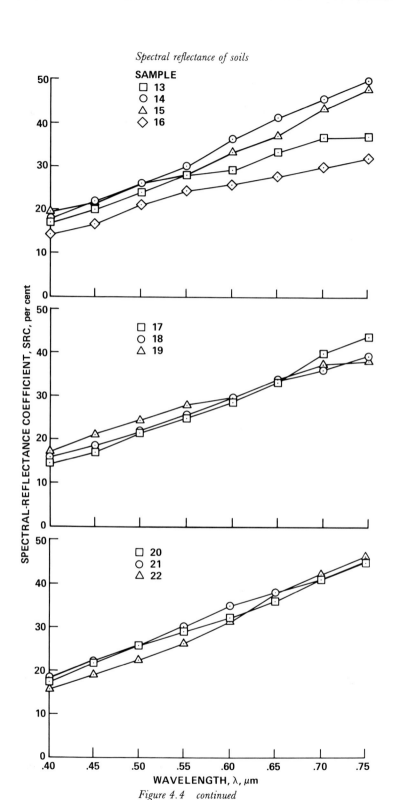

Figure 4.4 continued

drawn from laboratory studies to the field environment directly. To illustrate this, a comparison was made between field and laboratory measurements.

Field measurements of SBC were made with a field spectrometer for three soil types (chernozem, grey forest and peat–podzol) in the Ukraine. Only newly ploughed (but not wet) fields were analysed (Koltsov, 1975; Fedchenko and Kondratyev, 1981) and all measurements were made at a solar elevation of 45–50° at a time of clear skies. Soil samples were also acquired for laboratory measurements.

Laboratory measurements of the SBC were between 1·5 and 3 times higher than those measured in the field (Figure 4.5). To use the SBC of soils measured in the laboratory in the analysis of field measurements, the transfer function $K_\lambda^{(i)}$ must be determined using laboratory and field measurements of reflectance, r

$$r_{\text{field}} = K_\lambda^{(i)} r_{\text{lab}} \qquad (4.1)$$

The calculated values of $K_\lambda^{(i)}$ for the three types are given in Figure 4.6.

4.5. Spectral reflectance of soil as a function of its chemical and physical properties

Soil reflectance is a function of the soil's chemical and physical composition (Bowers and Hanks, 1965; Curran, 1985a). Although some investigators have used remotely sensed data recorded from aircraft and satellite altitudes (Fedchenko and Kondratyev, 1981), much work has been laboratory-based. This is partly because the latter allow the conditions for the formation of the reflected radiation field to be specified and the factors controlling the reflection characteristics of a soil to be assessed.

Soil can display a considerable range of colours (Bridges, 1978), and colour is therefore a useful indicator of soil type and soil properties. These may then be used to infer a variety of factors ranging from the soil's heat balance to its chemical composition. For instance, remotely sensed data have been used to assess salinity (Sharma and Bhargava, 1988), erosion potential (Pickup and Nelson, 1984; Pickup and Chewings, 1988) and soil type (Evans, 1972; White, 1977). However, deriving the relevant information is complex because a variety of factors determine soil spectral reflectance.

Many chemical and physical factors can influence a soil's spectral reflectance. From a remote sensing viewpoint the most important chemical parameters are the soil's humus, iron oxide, carbonate and soluble salt content, and the most important physical parameters are the soil's moisture content, texture, structure and morphology (Sadovnikov and Orlov, 1978; Curran, 1985a; Guyot et al., 1989).

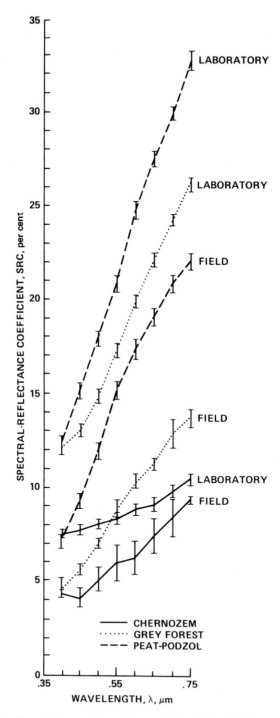

Figure 4.5 Spectral characteristics of three soils from field and laboratory measurements.

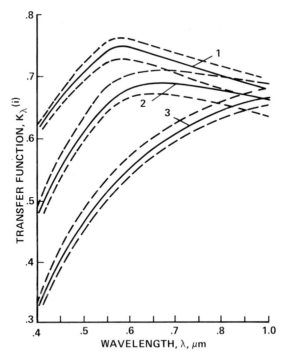

Figure 4.6 The transfer function $K_\lambda^{(i)}$ and its 70 per cent confidence intervals for three types of soil: (1) peat–podzol, (2) grey forest and (3) chernozem.

4.5.1. The effect of a soil's chemical properties on its spectral reflectance

The state and relative proportions of the soil's coloured chemical components and their horizontal and vertical distribution are responsible for a soil's colour and its spectral reflectance. Three components are primarily responsible for a soil's colour: humus, iron oxides and a variety of light-coloured substances (compounds of silicon and aluminium, calcium carbonates, etc.). Because these components exhibit considerably different spectral reflectances, their relative proportions will have a major influence on the spectral distribution of radiation reflected from a soil.

The humus content of a soil is a major determinant of its colour (Brady, 1984). It has a low reflectance and when in high concentrations is responsible for the grey-black colours of some soils (Orlov, 1966a; Orlov *et al.*, 1966, 1976; Orlov and Proshina, 1975; Sadovnikov and Orlov, 1978). Humus affects soil reflectance strongly in the $0\cdot70$–$0\cdot75$ μm wavelength range, but weakly in the $1\cdot00$–$2\cdot00$ μm range. The spectral reflectance of organic materials varies widely; for instance, pure humic acid has a very low total reflection (2–4 per cent when dry) because it is a strong absorber of light over the entire spectrum. Fulvic acids, on the other hand, display much higher reflection coefficients, especially in the green and red spectral regions, and have distinctly non-uniform reflectance spectra. Consequently, soils that are rich in humic acid tend to display a dark achromatic coloration, whereas soils with humus composed mainly of fulvic acids have a more saturated coloration.

A considerable amount of work has been undertaken in the USSR: (1) to establish correlations between soil humus content and spectral indices; (2) to understand the nature of light reflection by soils; and (3) to allow the extraction of soil properties from remotely sensed data. Empirically derived results show that soil reflectance depends on humus content and type. For instance, Orlov (1969) observed a direct linear dependence of reflectance on humic acid content (over a $0 \cdot 3 - 0 \cdot 8$ per cent content range) for a light brown forest soil in the Smolensk region. The total reflectance of the upper horizons of soils in Kazakhstan (independent of the carbonate content) was also determined to be dependent on the humus content and composition (Figure 4.7a). Spectral reflectance tended to increase with the transition from chernozem to brown to light brown soils, which was a negative relationship with humus content. For instance, the light coloration and relatively high total reflectance of greyish brown and light greyish brown soils resulted from their low, but fulvic-acid rich, humus content. Furthermore, the highest reflection coefficients were observed for the grey soils, which had the lowest humus content (Orlov et al., 1976).

Although correlations of the reflectance coefficients with the soil's content of humus, humic acid and fulvic acids were low ($r = -0 \cdot 39$, $-0 \cdot 38$ and $-0 \cdot 40$) they were nevertheless significant at the 95 per cent level of confidence. This is apparent even for soils of peculiar colouration. For instance, the subtropical soils in West Georgia display a colour which is strongly dependent on the rock type but which is modulated in a similar fashion by the organic matter content (Figure 4.7b; Orlov et al., 1978). However, a direct relationship between total humus content and reflectance for a wide variety of different soils has not been determined. This is primarily due to intersoil differences in the amount and spatial distribution of humus and the variability of its composition.

Iron compounds have a substantial effect on the spectral reflectance of soils (Vincent, 1973). The relationship between iron compounds and soil colour is more complex than it is for organic matter. Numerous iron compounds are found in soils, each displaying a different colour and spectral reflectance. The reflectance of an iron compound is strongly dependent on its degree of oxidation or hydration. In general iron protoxide gives the soil a bluish green colour, as is observed in gleyed and boggy soils (Curtis et al., 1976). Iron peroxide powder (Fe_3O_4) displays a low total reflectance with a minor peak in the blue-green spectral region. However, the reflected radiation from iron oxide (Fe_2O_3) has a much more heterogeneous spectral composition with an increase in spectral reflectance within the $0 \cdot 50 - 0 \cdot 58$ μm spectral interval; a feature which is also observed for ionized quartz. In remote sensing a high reflectance in the $0 \cdot 50 - 0 \cdot 54$ μm range is a characteristic feature of soils containing highly hydrated iron oxides and in humid climates such compounds can give the soil a

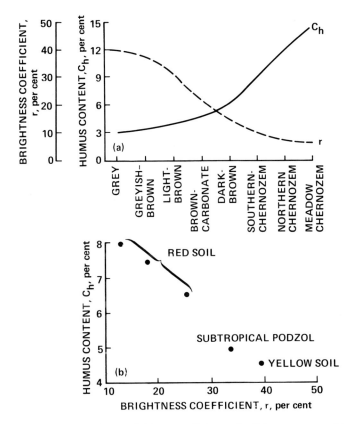

Figure 4.7 The relationship between brightness coefficient and soil humus content: (a) for the eight main soil types found in Kazakhstan, ordered from low to high humus content, and (b) for the five main soil types found in West Georgia.

yellow colour (Brady, 1984). Low and unhydrated iron oxides, however, display higher reflection coefficients in the $0 \cdot 50$–$0 \cdot 60$ μm spectral interval and are typically greyish brown or reddish in colour.

The presence and type of iron oxides found in a soil depends on a variety of factors. The soil moisture regime, initial composition of the silicate minerals in the soil-forming rock, the intensity of leaching and the character of the humus and its accumulation are all important variables. These influence the composition and distribution of the iron compounds which in turn produce a spectrally non-uniform reflection from the soil.

Although the influence of silicate iron, a component of the mineral crystal grids, on reflectance is relatively weak, that of non-silicate compounds is much stronger. A characteristic feature of reflectance from soils with non-silicate iron compounds is an increase in the spectral reflectance curve in the $0 \cdot 53$–$0 \cdot 60$ μm range. Orlov (1966b) showed that for artificial mixtures of iron oxide and quartz the dependence of reflectance on iron oxide contents of between 2 and 9 per cent could be expressed as

$$A_{0.64} = 8.5 - 4.9 [C_{Fe_2O_3}] \quad (4.2)$$

where $A_{0.64}$ is the reflectance coefficient at a wavelength of 0.64 µm and $C_{Fe_2O_3}$ is the percentage iron oxide content (Obukhov and Orlov, 1964). For some soils a close correlation between iron oxide content and spectral reflectance differences has been observed. For instance, the difference between the reflectance coefficients (or albedo A) at 0.65 and 0.40 µm has been shown to have a direct linear dependence on the iron oxide content. This can be expressed as

$$\Delta A = A_{0.65} - A_{0.40} = k + K[C_{Fe_2O_3}] \quad (4.3)$$

where k and K are soil-dependent constants.

These relationships have allowed the determination of iron content from a soil's reflection spectrum. For instance, in work by Orlov *et al.* (1966) the vertical distribution of mobile iron compounds estimated from spectral indices concurred with that determined by a conventional acetic-acid extract from both peat-podzol and greyish brown forest soils.

The influence of iron compounds on soil reflectance is most manifest in soils that are iron rich. The difference in reflectance coefficients at 0.65 and 0.40 µm (ΔA index) for West Georgian subtropical soils was positively related to the non-silicate iron content; $r = +0.90$. A regression of the non-silicate iron content over the 1.5–4.0 per cent range and the ΔA index (see Equation 4.3) was

$$\Delta A = 7.5 [C_{Fe}] - 1.5 \quad (4.4)$$

where $[C_{Fe}]$ is the percentage content of non-silicate iron in the soil. For soils with a non-silicate iron content outside this 1.5–4.0 per cent range more complicated relationships were apparent.

The dependence of spectrophotometric coefficients on the iron content alone can be determined if the soil samples are calcined at a temperature of about $900°C$. This destroys the organic matter and some clay mineral components of the soil, and results in the transformation of iron protoxides to iron oxides. Consequently, the colour and reflectance coefficient of calcined samples is determined by the iron oxide content and this dependence can be expressed as

$$\Delta A = 2.4 [C'_{Fe}] + 30.0 \quad (4.5)$$

where $[C'_{Fe}]$ is the percentage iron content of the soil.

The colour properties of a soil are also affected by the so-called light-coloured or white substances. The main members of this group are quartz, kaolinite, carbonates and some soluble salts. Additionally, some minerals (feldspar and calcite), fine-crystal gypsum and light-coloured aluminium compounds can give the soil a light coloration. On the whole these substances increase light reflection from soils and produce a more uniform spectral reflectance. Although these substances can sometimes be a major component of the soil, their influence on spectral reflectance is often secondary to that of humus and iron oxide. This is, in part, because the latter substances can form films on the surfaces of primary and secondary aluminosilicates. Nevertheless, the spectral reflectance of some soil horizons has been correlated with the content of light-coloured substances, notably soil carbonate content (Orlov and Proshina, 1975).

The spectral characteristics of a soil are a function of the spectral characteristics of its individual components and the content and distribution of these components. The total perception of the colour of a soil and its measurement with a spectrometer depends on the ratio of areas (volumes) of the coloured components on the given surface (volume) of soil. The combined effects of soil components on soil reflectance must be considered in order to determine quantitative techniques for estimating individual component contents from spectrophotometry. A number of equations have been developed to achieve this (Obukhov and Orlov, 1964; Orlov, 1966a; Orlov et al., 1966, 1976, 1978; Orlov and Proshina, 1975). These allow for different proportions of soil compounds, but typically assume them to be uniformly distributed spherical aggregates.

4.5.2. The effect of a soil's physical properties on its spectral reflectance

The intensity and spectral composition of radiation reflected from a soil also depend on the soil's physical properties. For instance, spectral indices have been shown to vary with the soil's aggregate size and moisture content (Orlov, 1969). However, the effect of aggregate size on the spectral distribution of reflected radiation is weak for aggregates larger than 2 mm. Similarly the total reflection observed from such aggregates is fairly constant, at least in the 2–10 mm range. Smaller aggregates vary widely in terms of the spectral composition of reflected radiation and the total reflection. For these the dependence of the reflectance coefficient (albedo) A and the aggregate size, $\tilde{d}$, have an exponential relationship (Orlov, 1966a):

$$A = 10^{-n\tilde{d}}k + R_\infty \qquad (4.6)$$

where k, n and R_∞ are constants characteristic of each soil type and horizon. Some examples of these constants are given in Table 4.1.

The effect of soil aggregation on reflectance can, as a first approximation, be estimated by comparing a soil's initial reflectance with that obtained after the soil has been ground. Grinding tends to increase the reflectance, but in a way that depends on the aggregate size. For relatively large aggregates (greater than approximately 0·5 mm) the chemical composition is not a major determinant of reflectance, with the reverse being true for relatively small aggregates. For aggregates of any size the major physical features affecting reflectance are the density of aggregate packing and surface roughness (Orlov, 1969; Orlov et al., 1976).

Table 4.1 Constants used in the Orlov (1966a) relationship between soil aggregate size and albedo.

Soil horizon	R_∞	n	k
Peat–podzol			
A_1	13·9	0·82	10·1
A_2	18·1	0·97	16·5
A_2B	19·6	0·92	24·4
B_1	17·0	0·89	13·0
A_2C	16·5	0·84	14·4
Usual chernozem			
A_1	5·6	1·58	4·8
Brown soil			
B_1	6·2	0·93	18·2
Saline soil			
B_1	6·8	0·44	12·8
Red soil			
A_1	11·3	1·14	6·5
B_1	14·4	0·36	4·1

Soil moisture content has a strong effect on soil reflectance. Wet soils display low SRCs and have a colour darker than similar dry soils. The relationship appears to be dependent, to some extent, on the soil's humus content. However, watering a soil to complete capillar water stress reduces the reflectance by a factor of 2–3, but does not markedly alter the shape of the spectral reflectance curve. Further wetting does not introduce a substantial change in reflectance unless the soil is wetted to or beyond its hydroscopic limit and the soil particles are covered by a film of water. In this situation and at certain scene–sensor geometries specular reflection may cause the overall soil reflectance to increase. Note also that the water will be exchanged for air in the soil pores. This will lead to a chemically reducing environment and, amongst other factors, will change the proportions of ferric and ferrous iron oxide. Consequently, the effect of soil moisture content has other related effects on soil reflectance. In general, however, soil reflectance is negatively related to soil moisture content, although this relationship can break down when the soil is wetted beyond a critical level (Neema et al., 1987).

The spectral proportions of radiation reflected by humus-accumulating horizons of soils has been shown by Sadovnikov and Orlov (1978) to vary little with wetting, except for yellow and red soils. For a yellow soil the difference in reflectance coefficients at wavelengths of 0·64 and 0·45 μm for air-dried soil was 22·8 per cent, but was only 10·1 per cent for a soil with a 20 per cent moisture content, although this may also be a function of the varying hydration of the soil's iron oxides. Soil moisture has its largest effect on the character of the spectral-reflectance curves in the near-infrared. A strong decrease in reflectance was observed in the 1·40–1·50 μm wavelength interval, associated with the

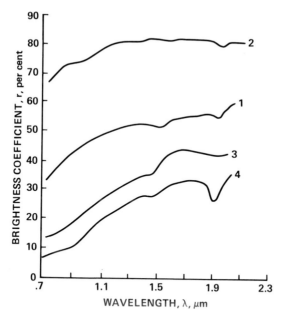

Figure 4.8 Soil spectral reflectance (per cent) in the infrared. (1) Peaty soil, 0–12 cm depth; (2) Peaty soil, >12 cm depth; (3) Typical chernozem, 0–50 cm depth; (4) same as (3) but wetted to a moisture content of 28·6 per cent.

water absorption band. Moisture was also responsible for a decrease in reflectance in the 1·80–1·90 μm interval (Figure 4.8). Increasing the moisture content of a wet soil results in only a minor change in reflectance. Also, for at least three soil types (peat–podzol, chernozem and red soils) there was no significant difference in the reflectances of samples observed at 0 per cent moisture (oven-dried) to those that were at the hygroscopic limit (air-dried). Consequently, Sadovnikov and Orlov (1978) concluded that the diffuse reflection spectra of dry soil can be measured without the need for oven-drying.

Between the hygroscopic limit and complete capillar water capacity the SRCs were found to be negatively related to soil moisture content. For the soils studied, the dependence of total reflection on soil moisture content in the 5–20 per cent content range could be expressed in terms of the regression equation

$$A_\Sigma = B_1 - B_2 \omega \tag{4.7}$$

where ω is the percentage soil moisture content. Examples of the B_1 and B_2 terms are given in Table 4.2. The K_i values define the range of the relationship, and are equal to the ratio of the total SRC for air-dried soils to that for soils in a capillar-saturated state, varied from 1·4 in peat–podzol and brown soils to 2·1 in yellow soils.

Table 4.2 *Constants used in the relationship between soil moisture content (in the range 5–20 per cent) and albedo.*

Soil	B_1	B_2	$K_{i_{\min}}$	$K_{i_{\max}}$
Peat–podzol	16·0–18·6	0·25–0·61	1·4	1·8
Grey forest	20·1–26·8	0·34–0·57	1·8	1·9
Chernozem	7·2–11·1	0·16–0·22	1·6	1·7
Brown	12·2–14·1	0·09–0·17	1·4	1·5
Saline	25·3	0·58	1·2	
Red	18·1–31·2	0·70–0·92	1·2	1·5
Yellow	16·9	0·61	2·0	2·1

4.6. Concluding comments on the reflectance of soils in the laboratory and field

The total and spectral reflectances of soil are a complicated function of the composition and structure of its surface layer. The reflectance coefficients depend largely on the humus, iron oxide(s) and moisture contents, as well as the size and composition of the aggregates. The main features can be summarised as follows:

(1) The colour of a soil is determined primarily by its chemical composition. Humus compounds give a characteristic dark coloration in the visible and near-infrared spectral regions. Iron oxides of different degrees of hydration or oxidation give a characteristic selective coloration in the red wavelengths. Silicon and aluminium compounds and calcium carbonates give the soil light colours.
(2) Organic matter affects the reflectance of soils primarily in the $0·70–0·75$ μm wavelength interval. Reflectance coefficients are, in general, negatively related to the organic matter content.
(3) The spectral composition of radiation reflected by soil with aggregates larger than 2 mm is independent of the size of the aggregates. The spectral distribution of reflected radiation from a soil will be different when comparing samples of the soil above and below this threshold size.
(4) Soil reflectance generally decreases with increasing soil moisture; this decrease is especially large for soils low in organic matter, where wetting a soil to its capillar water capacity can reduce its reflectance by a factor of 2–3. However, the shape of the spectral reflectance curve remains relatively constant.
(5) Before the laws of radiation reflection from soils can be used in the analysis of remotely sensed data, account must be taken of the natural variability of soils (Beckett and Webster, 1971) and the atmosphere.

Some of these features are illustrated further in Chapters 5 and 11.

5
Remote sensing of soil humus

Humus is a complex aggregate of amorphous, dark-coloured substances (Vaksman, 1937; Brady, 1984) formed from the decomposition of plant and, to a lesser extent, animal remains in the soil. Decomposition is influenced by a variety of factors, notably vegetation cover, micro-organism activity and the physical and chemical properties of the soil. For instance, humus formation is relatively inactive in desert and semi-desert areas, but is active in areas of meadow and steppe vegetation (Tiurin and Konova, 1962).

Chemically, humus is a very complex mixture of compounds dominated by humic acids, fulvic acids and humins (Russell, 1973; Trudgill, 1988). Humic acids are a group of high-molar compounds with common properties and composition. Their molecules have a loose, spongy composition with many internal pores (Konova, 1963).

Fulvic acids differ from humic acids in that they have a higher oxygen and lower carbon and nitrogen content (less than 55 per cent; Russell, 1973). They play an active role in the decomposition of rocks and minerals, and are often very soluble in water, alcohol, alkali and mineral acids. Because of the variable colloid-chemical composition of fulvic acids, they can be transformed into humic acids and vice versa. Humins are humic acids that have been modified chemically by the mineral component of the soil (Tiurin and Konova, 1962). As such, they have properties similar to those of humic rather than fulvic acids.

The composition of soil humus depends largely on the local environmental conditions, notably temperature and precipitation (Table 5.1; Tiurin and Konova, 1962; Russell, 1973). For instance, a moderate precipitation regime favours the formation of humic acids and wetter regimes favour the formation of fulvic acids.

Dergacheva and Kuz'mina (1978) measured the absorption spectra of humic and fulvic acids in the ultraviolet and visible spectral range (Figure 5.1). The humic acid spectra had absorption minima in the $0 \cdot 260-0 \cdot 275$ μm waveband and maxima in the $0 \cdot 225-0 \cdot 230$ and $0 \cdot 260-0 \cdot 275$ μm wavebands. In general the humic acid absorbed radiation more strongly than the fulvic acid, this difference being most clear in the shorter wavelengths (Dergacheva, 1972). The

Table 5.1 *The chemical composition and environment of formation for humus in eight soils from the USSR and one soil (red) from China (Tiurin and Konova, 1962.)*

Soil	Humus content (per cent)	Acids in soil humus		Humic/fulvic	Annual precipitation (mm)	Mean temperature maximum/minimum (°C)
		Humic (Relative units)	Fulvic			
Podzol (heavy podzol and peat–podzol)	2·5–4·0	12–20	25–28	0·6–0·8	500–600	15–18/–10
Grey forest	4·0–6·0	25–30	25–27	1·0	500–550	18/–10
Thick and standard chernozem	7·0–10·0	35–40	15–20	1·5–2·5	400–450	20/–10
Dark brown	3·0–4·0	30–35	20	1·5–1·7	300–350	24/–13
Greyish brown desert-steppe	1·0–1·2	15–18	20–25	0·5–0·7	200–250	25/–10
Typical grey	1·5–2·0	20–30	25–30	0·8–1·0	350	27/–1·5
Chernozem	4·0–6·0	15–20	22–28	0·6–0·8	2400	22/6
Red	3·0	12	30	0·4	1100	20/9
Lateritic	4·0	6	34	0·2	1900	28·9/15·5

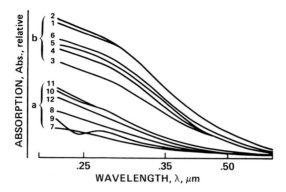

Figure 5.1 Light absorption spectra of humic and fulric acids. Humic acids: (1) Chestnut soil, A horizon; (2) Salty chestnut soil, A_1 horizon; (3) Salty chestnut soil, B horizon; (4) Chernozem soil, A horizon; (5) Salty chernozem soil, A horizon; (6) Salty chernozem soil, B horizon. Fulvic acids: (7) Chernozem, A horizon; (8) Salty chernozem, A_1 horizon; (9) Salty chernozem, B horizon; (10) Chestnut soil, A horizon; (11) Salty chestnut soil, A_1 horizon; (12) Salty chestnut soil, B horizon.

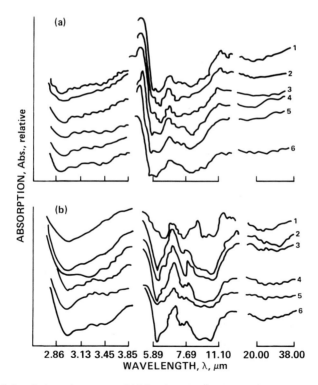

Figure 5.2 Infrared absorption spectra of (a) humic and (b) fulvic acids. (1) Southern chernozem, A horizon; (2) Salty chernozem, A horizon; (3) Salty chernozem, B_1 horizon; (4) Chestnut soil, A horizon; (5) Salty chestnut soil, A_1 horizon; (6) Salty chestnut soil, B_1 horizon.

infrared absorption spectra of humic acids (Figure 5.2) showed strong absorption in the 2·86–3·03 μm waveband (Dergacheva and Kuz'mina, 1978). This was due to oscillations of OH groups and free and bounded hydrogen bonds of the NH groups. In addition, narrow-band absorption was observed within the 3·49–3·57 μm waveband, and in two wavebands centred on 3·40 and 7·29 μm. Fulvic acids had a similar infrared absorption spectra with the addition of a narrow absorption band centred on a wavelength of 3.13 μm. The ultraviolet to infrared absorption spectra of humic and fulvic acids were therefore similar in form, and differed primarily in the intensity of the absorption bands (Figures 5.1 and 5.2).

The reflection spectra of humic acids and fulvic acids were, by contrast, dissimilar (Obukhov and Orlov, 1964; Tolchelnikov, 1974). Humic acids are strong absorbers of light and so have low reflectance. For example, the total light reflected from a humic acid (extracted from a thick chernozem soil) was approximately 2 per cent. Furthermore, powdery humic acid is practically an achromatic substance (Obukhov and Orlov, 1964). However, fulvic acids are more-selective reflectors; they display reflectance peaks in green and red wavelengths.

5.1. The reflectance properties of soil humus and its application

In recent years the key role played by soil humus in soil formation and fertility has become more widely accepted. For example, variables such as the gross humus content and its vertical distribution in the soil are now widely used in the USSR as diagnostic indicators for the genetic classification of soils. Several reliable field and laboratory techniques exist for determining the characteristics and amount of humus in soil (Orlov and Grishina, 1981). Although these techniques are reliable, their application has been limited because they are cumbersome and time-consuming. Consequently, recent research has aimed at developing remote sensing techniques for assessing humus content.

The major soil characteristic which can be analysed remotely is spectral reflectance. Attempts to establish relationships between the humus content of a soil and its spectral reflectance can be traced back to the work of Pokrovsky (1929). However, the optical approaches developed were not used extensively until the 1950s, when reliable techniques and instrumentation became available. These investigations revealed that such soil componds as humus and iron oxides had major effects on the spectral reflectance of soils (Orlov, 1960, 1966b; Orlov *et al.*, 1966, 1976; Tolchelnikov, 1974; Orlov and Grishina, 1981). For instance, soils in the subtropical Lenkoran lowland (Azerbaidjian) displayed a high correlation between a reflectance coefficient measured at 0·72 μm and soil humus content (Zyrin and Kuliev, 1967).

Similar results were then obtained for other soils using a variety of reflectance measures. Mikhailova (1970), for instance, used such variables as total reflectance, tangent of the slope of the reflectance curve and ratioed reflection coefficients. Sorokina (1967) also used total reflection and found it to have a correlation coefficient of around $0 \cdot 9$ with humus content for soils with up to 5 per cent humus. Furthermore, by taking account of the effects of depth and carbonates on the relationship, the observed correlation coefficient could be increased to around 0.98.

Relationships between soil humus content and spectral reflectance provide valuable information for many applications. Aside from the information on soil humus, other features that are related to soil humus can be inferred from spectral reflectance. For instance, Latz et al. (1981) used measurements of the SBC to analyse the possibility of recognizing the degree of erosion-induced reduction in soil productivity in Indiana, USA. In their investigation SBCs were measured in the laboratory over the $0 \cdot 52 - 2.32$ μm spectral range, with a wavelength step of 10 nm. The soil samples were obtained from sites subject to water erosion, and a simple classification of the degree of erosion was based on the amount of the soil's A horizon that had been lost. The soil samples were chopped, dried and wetted to a reference moisture content, and were characterised in terms of their particle-size distribution, organic carbon content, iron oxide content and cation exchange capacity. Three useful reflectance categories could be identified.

(1) Soils with a high content of humus, characteristic of non-eroded mollisol-type soils.
(2) Soils with a lower humus content and good drainage, characteristic of eroded mollisol-type soils and non-eroded alfisols and utisol soils.
(3) Soils with a low humus content and a relatively high iron-oxide content, characteristic of utisol and eroded alfisol soils.

Humus and iron oxide were therefore the main determinants of the SBC, with iron oxide being important only at low humus contents. These three reflectance categories and, therefore, soil erodability were most reliably assessed from SBC data at approximately $0 \cdot 8$ μm, which points to the value of band 4 of the Landsat Thematic Mapper (TM) for surveying soil erodability.

Information on the angular distribution of the soil's SBC is required if remotely sensed data are to be used to derive data on the physicochemical properties of soils and to map them accurately. Laboratory-based studies on the angular distribution of SBC have shown that some soil types can be easily identified (Stoner and Baumgardner, 1980; Stoner et al., 1980). Soils with similar colour characteristics may, for example, have quite different SBC curves, especially in the near-infrared region. Additionally, regionally averaged SBC curves revealed that SBC tended to increase as temperature increased and precipitation decreased as a result of reduced humus content. A series of correlation analyses between spectral reflectance and organic matter content revealed a negative correlation with the logarithm of the organic matter content $(r = -0 \cdot 78)$ over the entire $0 \cdot 52 - 2 \cdot 32$ μm range. In the $2 \cdot 08 - 2 \cdot 32$ μm range, negative correlations with moisture content $(r = -0.95)$, clay content $(r = -0 \cdot 90)$, cation-exchange capacity $(r = -0 \cdot 84)$ and iron content

($r = -0.70$) and a positive correlation with fine and moderate sands were observed. However, it is unlikely that such parameters can be estimated reliably from SBC data over a wide range of environments as climatic conditions, parent rock and drainage, amongst other environmental factors, will modulate the SBC. However, within a defined climatic zone, regression analyses have shown that SBC can be used to estimate reliably the soil's humus content, particle-size distribution, moisture content, iron oxide content and cation-exchange capacity.

The results of these and similar studies have allowed the definition of spectral intervals that are likely to be useful in recognising certain soil properties. These are: $0.52-0.62$ μm (humus), 0.70 and 0.90 μm (iron), 1.00 μm (iron and hydroxyl) and $2.08-2.32$ μm (soil moisture). Additionally, the strong reflection bands at $1.22-1.32$, $1.55-1.75$ μm and longer wavelengths may contain information on a variety of soil properties that have yet to be characterised. Given these spectral intervals, some useful information on soil properties may potentially be derived from the Landsat TM sensor using band 2 ($0.52-0.60$ μm), band 4 ($0.76-0.90$ μm), band 5 ($1.50-1.75$ μm) and band 7 ($2.08-2.35$ μm).

Using these relationships, soil reflectance data can be used to estimate soil properties for use in further inference. For example, the potential of erosion and its implication for land use have been derived from Landsat Multispectral Scanning System (MSS) imagery (Chaume and Phu, 1980). Using a principal component analysis Mather (1987) revealed that only the $0.50-0.60$ μm and $0.60-0.70$ μm wavebands were required for analysing soil erodability, as they accounted for 95 per cent of the variance in soil erosion. This is because these bands contain most of the information on soil humus content, which is a major variable in determining soil erodability.

Similarly, Steglik (1982) was able to identify eroded and non-eroded soils from multispectral aerial photography. Maximum differences in the optical densities between the eroded and non-eroded soils were observed on photographs filtered to record spectral intervals centred at 0.55, 0.62 and 0.70 μm (interval widths were 0.04 μm). Relationships between reflectance and soil factors such as clay content, organic carbon content and productivity (Schubert et al., 1980) have also been reported widely in the USSR.

Clearly remote-sensing systems operating in the visible and near-infrared wavelengths have considerable potential for soil classification and mapping (Evans, 1972; Johanssen and Dacosta, 1980). For instance, a comparison of soil maps of the USSR produced from Landsat MSS imagery at a scale of 1 : 250 000 with those produced by conventional techniques revealed that those derived from the remotely sensed data were often more detailed and accurate.

In conclusion, therefore, a number of researchers have shown that soil humus content and spectral reflectance are correlated. This correlation can be observed with SBC measurements at one (Zyrin and Kuliev, 1967), two (Mikhailova, 1970) or more wavelengths (Latz et al., 1981), or even from the total reflectance curve (Sorokina, 1967; Tolchelnikov, 1974). However, it is unlikely that there is a single wavelength or point on the spectral reflectance curve which could be used to provide consistently a reliable estimate of soil humus content. The wide variety of soil and environmental parameters that affect reflectance make the entire visible spectral reflectance, normalised against the spectral composition of the incident radiation, a more reliable measurement for humus estimation. However, although some Soviet researchers have used the entire reflectance curve (Sorokina, 1967; Tolchelnikov, 1974), few have normalised the data against the spectral composition of incident radiation.

5.2. Theoretical studies of the relationship between soil spectral reflectance and humus content

Spectral reflectance is the principal parameter used in the remote sensing of soils. However, the interactions of radiation with a soil are very complex, as individual soil components vary in their ability to scatter and absorb radiation. Furthermore, the manner in which these components are mixed and the effect of multiple scattering confound attempts to determine the relationships between reflectance and soil parameters. Empirical solutions are often sought as a result of the soil's complex composition and the problems of describing mathematically the interactions between radiation and soil properties. For the estimation of humus content from remotely sensed data the initial stage may be to relate reflectance, r, to the humus content, C_h (Pokrovsky, 1929):

$$r = c_1 e^{-c_3 C_h} + c_2 \tag{5.1}$$

where c_1, c_2 and c_3 are empirically derived coefficients related to the reflectance of soil-forming rock, the reflectance of humus and a measure of the soil particle-size

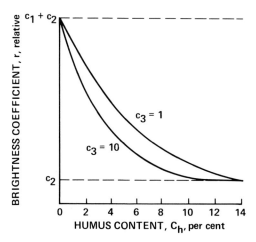

Figure 5.3 The relationship between the brightness coefficient and soil humus content for different values of the reflectance coefficients, c_1 and c_3.

distribution, respectively. The dependence of r on C_h derived for a range of different values of c_1, c_2 and c_3 are shown in Figure 5.3; the upper broken line ($r = c_1 + c_2$) represents fine grain sizes ($c_3 \to \alpha$). Empirically derived relationships such as these allow the effect of particle size on spectral reflectance to be determined for a variety of soil types. Figure 5.4 shows the spectral reflectances of soils with no, low and high humus contents as a result of three particle-size distributions. The effects of particle-size distribution on soil reflectance were more substantial for the soils with low than for those with high humus contents, and must therefore be accounted for in any attempt to estimate soil humus content using remotely sensed data.

To use this relationship between soil reflectance and humus content, it is necessary to measure and express soil reflectance in terms of brightness or brightness coefficients. For humus contents of up to 6 per cent the linear approximation

$$W_{shs} = \rho_1 W_h + (1 - \rho_1) W_s \qquad (5.2)$$

may be used, where W_{shs}, W_s and W_h are the parameters describing the reflectance of the soil-forming rock–humus system, the soil-forming rock and the humus, respectively, and ρ_1 is the weighting coefficient for humus content in the soil-forming rock–humus system.

A variety of different variables which characterise quantitatively the spectral reflectance of soils can serve as W-parameters (Equation 5.2). For instance SBC, SRC, total reflectance and the sums of the colour coordinates may be used. As an example, colour coordinates and their sums calculated for the loess soils in Figure 4.4 are given in Figure 5.5 (pp. 92–93). The techniques used for colour-coordinate calculation are described in the literature (Fedchenko, 1982a,b, 1983;

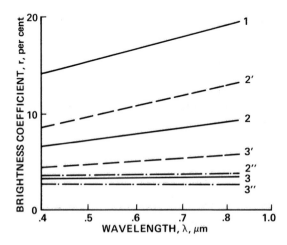

Figure 5.4 Spectral characteristics of soil humus horizons: (1) no humus content, (2) low humus content and (3) high humus content. A prime denotes compressed soil and a double prime ploughed soil.

Kondratyev and Fedchenko, 1981a, 1982e,f; Kondratyev *et al.*, 1982b) and in Section 2.4.

To determine ρ_1 from Equation 5.2 it is necessary to know W_{shs}, W_s and W_h. Remote sensing allows only the measurement of W_{shs}; consequently, prior knowledge of W_s and W_h is required.

5.3. Laboratory studies of the relationship between soil colour coordinates and humus content

Samples of three soil-forming rocks were analysed in the laboratory to determine a quantitative relationship between humus content and colour coordinates. The rocks were all loesses, but they had different mechanical compositions. The reflectance properties of nine samples of known weight (10–12 g) were measured with an SF-18 spectrophotometer. For each sample a small proportion (1·5–3 g) of soil with a known and high humus content was added to the rock. The rock and soil were mixed together and weighed, and the spectral analysis was repeated.

Consequently, samples of a variety of soil and rock proportions were analysed, ranging from pure rock to a mixture with equal proportions of soil and rock. Finally, the entire procedure was repeated in reverse; rock was added to samples of the same soil. The resultant spectra allowed the colour coordinates to be calculated with respect to a standard emission source (for spectral intervals $\Delta\lambda = 0\cdot01$ μm; Equation 2.9). Then, the sum of the three coordinates was found

$$W = X + Y + Z \tag{5.3}$$

A nearly-linear correlation was observed between W and soil humus content in the 0–7% range (Figure 5.6). This relationship could then be used to estimate the humus content of other soil samples from their W value.

To illustrate the accuracy of this procedure, 32 soil samples of different humus content were analysed. These were air-dried, chopped and sieved, and their spectral reflectance was measured with an SF-18 spectrophotometer. Colour coordinates, and thus W, were derived from the resultant spectral reflectance curves. The humus content was determined using both a chemical analysis and W and the relationship in Figure 5.6 (Figure 5.7). In most cases the difference between the two was less than 4% but it reached almost 10% for samples with low humus content.

5.4. Laboratory studies on the effect of soil-forming rock on the relationship between soil colour coordinates and humus content

The results obtained in Section 5.3 indicate that there is a strong negative correlation between the colour coordinates of a soil and its humus content.

Spectral reflectance of soils

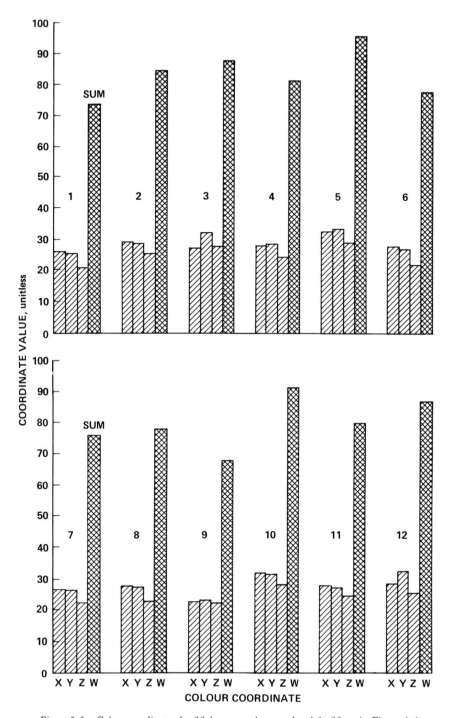

Figure 5.5 Colour coordinates for 22 loess samples, numbered 1–22 as in Figure 4.4.

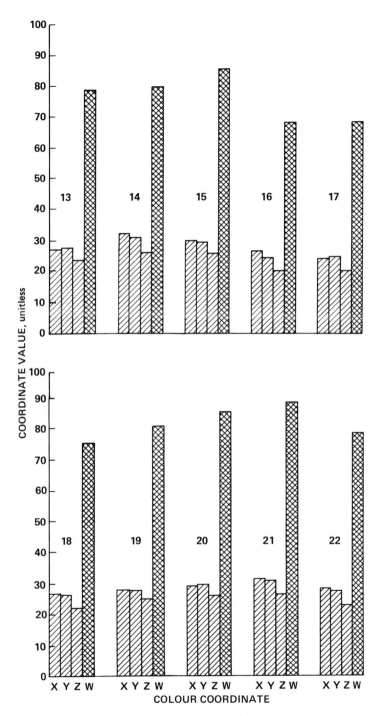

Figure 5.5 continued

Spectral reflectance of soils

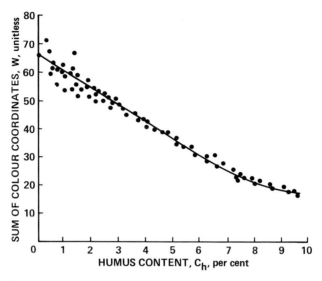

Figure 5.6 The sum of soil colour coordinates as a function of humus content (laboratory study).

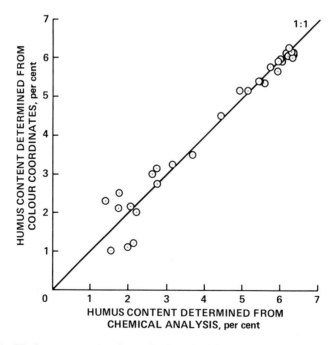

Figure 5.7 The humus content in soil samples determined from the sum of the colour coordinates and from conventional chemical analysis.

However, this is too general a relationship for reliably determining soil humus content from remotely sensed data. It is, for instance, not always possible to determine even the relative order of humus content for soils on which only the colour coordinates are known (e.g. Figure 5.8). Thus, using these reflectance curves to calculate colour coordinates and estimate soil humus content will be inaccurate, because the spectral reflectance of a humus-containing soil is also strongly influenced by the reflectance of the soil-forming rock (W_s). To assess the importance of W_s on the calculation of ρ_1 from Equation 5.2, a simple three-stage experiment was performed.

Stage 1: deriving a calibration between the colour coordinates and soil humus content. A soil-forming loess from the Kiev region (number 10 in Figure 5.5) was dried and sieved. As previously, the sample was weighed and its reflectance was measured using an SF-18 spectrophotometer. Small portions of soil with a known and high humus content were then mixed with the loess and its reflectance was again measured. This procedure was continued to a point where the sample was composed of equal weights of soil and loess. As before, this experiment was repeated in reverse by adding loess to the soil. A series of spectra were measured over this wide range of humus contents, the colour coordinates were estimated from these spectra and W was determined. Again there was a strong negative correlation between W and humus content (Figure 5.9).

Stage 2: the relationship between the colour coordinates and soil humus content for one loess. A sample of soil-forming loess from the Poltava region (sample number 14 in

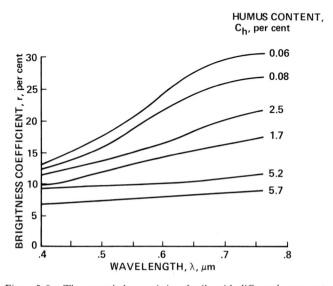

Figure 5.8 *The spectral characteristics of soils with different humus contents.*

Spectral reflectance of soils

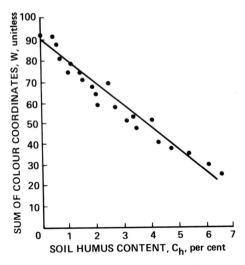

Figure 5.9 The sum of soil colour coordinates as a function of humus content (laboratory study).

Figure 5.5), which has a similar SBC to the Kiev soil, was treated in a similar manner to the soil in stage 1. Eighteen small portions of the soil with a known and high humus content were added to it, the reflectance was measured, W was derived and the soil humus content was estimated using the relationship depicted in Figure 5.9. Results are shown in Figure 5.10a.

Stage 3: the relationship between the colour coordinates and soil humus content for several loesses. Twenty samples of soil with a known and high humus content, but of different weights, were mixed with one of 22 soil-forming loess samples, the spectral reflectances of which are given in Figure 5.5. The reflectance was measured, W was derived and the soil humus content was estimated using the relationship in Figure 5.9. Results are shown in Figure 5.10b.

Figure 5.10 indicates that the accuracy with which humus content can be estimated from soil spectral reflectance depends on the spectral reflectance of the soil-forming rock, which in this case was a loess. Accurate estimates may be obtained when the spectral reflectances of the soil-forming rock used in forming the calibration (e.g. Figure 5.9) and that of the sample for which the humus content is to be estimated are similar. If the spectral reflectances of these two differ markedly, then large errors in estimating humus content are likely. The humus content was important because at low concentrations (1–2 per cent) the soil-forming rock had a strong influence on the soil reflectance, so estimation was inaccurate, with errors of around 200–300 per cent (Figure 5.10b). Alternatively, when the sample's humus content was greater than about 2.5 per cent the soil-forming rock had little influence on the estimate's accuracy.

A soil's reflectance is therefore a function of both its humus content and the reflectance of soil-forming rock. Because this complicates the remote estimation

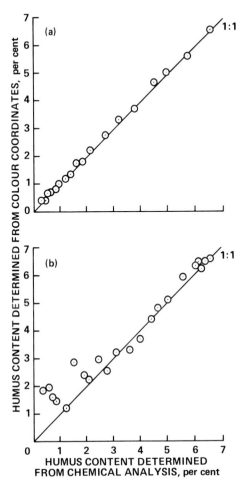

Figure 5.10 The humus content of soil samples, determined from the sum of the colour coordinates and from conventional chemical analysis for (a) one loess and (b) several loesses.

of soil humus content, attempts have been made to identify an optimal threshold humus content beyond which the soil-forming rock has a negligible influence on the relationship between colour coordinates and humus content. To achieve this, four soil-forming rocks (loess) of different spectral reflectances (numbers 2,5,6 and 9 in Figure 5.5) were mixed with different proportions of soil with a high humus content. Their reflectance properties were measured and the humus content estimated from the sums of the colour coordinates.

The results (Figure 5.11) show that, as before, the soil-forming rock does significantly affect the relationship. However, for humus contents greater than about 2·5 per cent the effect is negligible. Therefore, the humus content of a soil can be estimated accurately from Equation 5.2 with known values of W_{shs}, W_s

Spectral reflectance of soils

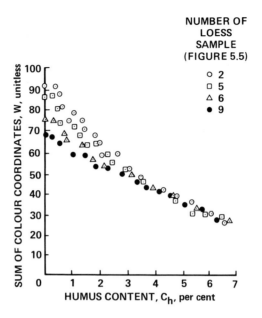

Figure 5.11 The relationship between the sum of the colour coordinates and humus content of soils formed on four different loesses.

and W_h provided the sample's humus content exceeds 2·5 per cent or the rock from which the soil has formed is known.

5.5. Field studies of the relationship between soil colour coordinates and humus content

Although laboratory measurements have shown that there is a strong correlation between colour coordinates and soil humus content, ground-based field measurements are typically used in the USSR to determine the calibration curve for use with remote-sensing investigations. This is necessary because laboratory measurements of SBC are typically 1·5–3 times higher than those measured in the field (Section 4.4). This is because in the field the measured SBC (not SRC as in the laboratory) is dependent on such features as the viewing geometry of the sensor and illumination source (Rachkulik and Sitnikova, 1981). In a study of soil reflectance in the Central Ukraine 30 sample sites of 50 cm × 50 cm were used to derive a calibration curve between colour coordinates and soil humus content. The sample sites were in freshly ploughed fields and spanned the entire range of humus contents for the region.

At each sample site measurements were made with a field spectrometer (Koltsov, 1975) mounted 1·5 m above the soil surface. Attempts were made to control the factors affecting soil spectral reflectance, such as its moisture content and conditions of cultivation and illumination. Consequently, only soils which had not recently been subject to precipitation were selected (Kondratyev and Fedchenko, 1980d). Additionally, the character of the soil surface cultivation was similar for each sample and the maximum aggregate size was approximately 6 cm. Finally, all measurements were made under clear-sky conditions at solar elevations of $40-45°$.

Because the SBC of natural surfaces depends on the ratio between scattered (S) and direct (D) solar radiation, this ratio was measured at the time of each SBC measurement. As the maximum difference between D/S ratios did not exceed 15 per cent, it was unlikely that it had a significant effect on the SBC measurements.

The humus content was then estimated in the laboratory using the procedure outlined above. The surface layer of the soil, which gives the reflectance in the field, was sampled immediately after each radiometric measurement, and these samples were chopped and sieved. Three subsamples, each of approximately 15-20 g, were taken from each sample site and their reflectance was measured with an SF-18 spectrophotometer. If the spectral reflectance curves for the three samples at a site differed by more than 10 per cent they were remixed and their reflectance was again measured.

Using these data a calibration curve between the sum of the field-measured colour coordinates (y) and humus content (x) was drawn (Figure 5.9). This graph was then used to estimate the humus content of soils from airborne measurements of SBC on the assumption that the field-measured colour coordinates had little error and the form of the relationship for x on y was the same as that for y on x (Curran and Hay, 1986).

5.6. Aircraft studies of the relationship between soil colour coordinates and humus content

During July and August 1973, airborne spectrophotometric measurements were made for preselected sites in the Ukraine and Moldavia. These sites lay approximately 5-10 km apart along flight tracks that were 30-40 km apart. The aircraft flew directly over each sample field, spectrometric measurements were made and the flight height, cloudiness and illumination conditions were noted. The flight height was varied according to the field size. It averaged approximately 100 m, at which the field of view of the instrument was 38 m x 40 m. All measurements were made under clear or partially cloudy skies. The solar elevation, derived from tabulated figures for known latitude, season and time, was never below $40°$, so there was no need to correct for non-orthotropicity of the reference material.

The accuracy with which soils can be mapped on the basis of their spectral reflectances depends on the timing of data collection. Previous research has shown that soils in the Ukraine and Moldavia are best studied during the ploughing season (July and August), when most of the territory is free of crops.

The soils are cultivated in a similar way, so 2–3 days after ploughing the effect of soil moisture on SBC variability is minimized. Furthermore, during this season there are a comparatively large number of sunny days and, consequently, atmospheric influence on the SBC is low.

Data were collected using a fast-operating spectrometer (Koltsov, 1975) onboard an AN-2 light aircraft. The reflected radiation from a reference screen and soil target were recorded on a light-sensitive film in the form of oscillograms. The conversion of spectral brightness values on these to SBC was made by measuring the ordinates of the reference and the soil target, the zero line being assumed as a starting point. The dependence of the soil SBC on solar elevation was excluded from SBC calculations.

Although the monotonic positive relationship between spectral reflectance and wavelength is often observed for soils in the laboratory and field, it is not always so when using an airborne spectrometer. The values derived from the latter can depend on the spectrometer's speed of operation. In this case the spectrometer recorded soil spectra in 5 s and the aircraft's speed was approximately 50 m s^{-1}, so the spectrometer would have moved 250 m while recording one soil spectrum. The spectra are therefore for a long and linear target, the homogeneity of which will determine the accuracy and reliability of the resultant spectral reflectance data.

For some soils the SBCs from which the spectral reflectance curves were drawn were variable, and they were smoothed with Chebyshev's polynomial before the spectrograms were estimated quantitatively. Fortunately, this was only necessary for approximately 3 per cent of the data. The colour coordinates were then determined and summed, and this information, in conjunction with the calibration curve (Figure 5.12), was used to estimate the percentage humus content for the sites at which SBC was measured.

The final stage was to map the percentage humus content of the Ukrainian and Moldavian soils. A section of this, for the Kharkov region, is illustrated in Figure 5.13. A comparison of this map with other soil maps shows that for those

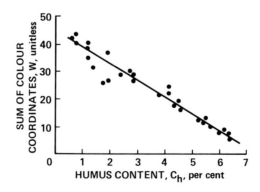

Figure 5.12 The sum of soil colour coordinates as a function of humus content (field measurements).

regions with a high humus content in their soils (such as Kharkov, Voroshilovgrad, Poltava, Kirovograd, Odessa, Dnepropetrovsk and Donetsk) there is a high level of agreement. In other regions less coincidence was observed; for example, a substantial difference was apparent for the Polesye region, which is composed of a different soil-forming rock (Section 5.4). Overall, however, the technique did allow the estimation and mapping of soil humus content to a moderate level of accuracy.

In autumn 1982 this experiment was repeated in order to test the technique. The spectral reflectance of soils in the Maloyaroslavets district of the Kaluga region were obtained about the time that the USSR Central Institute for Agrochemical Service (CIAS) was collecting soil samples to assess humus content by conventional methods. It was therefore possible to compare the humus

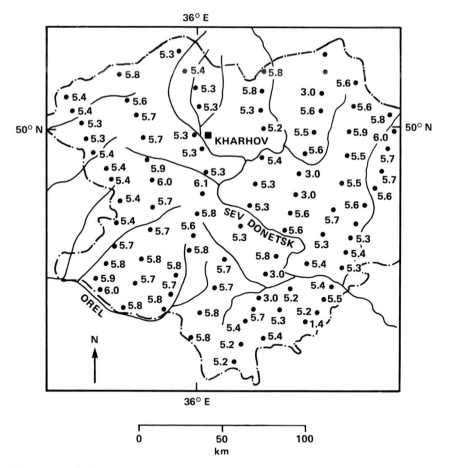

Figure 5.13 A schematic map of the humus content of soils in the Kharkov region of the USSR. The numbers are the estimated per cent humus content at the points of radiometric measurements.

content estimate obtained by the CIAS with that derived from the spectral reflectance data.

In general, the two estimates agreed, especially for soils of medium to high humus content. Although some of the deviation from the CIAS estimate will be a function of the colour-coordinate technique, other factors must be considered. For instance, the CIAS data refer to point estimates of the humus content, whereas the remotely sensed data are effectively an average of the humus content of the entire field. A common conclusion to such work is that the remotely sensed data are not as accurate as *in situ* measurements for point data, but allow information for very large areas to be obtained quickly.

5.7. Concluding comments on the remote sensing of soil humus

Laboratory, field and airborne sensor measurements have demonstrated that: (1) the spectral reflectance of a soil is negatively related to its humus content; and (2) this relationship and further measures of spectral reflectance can be used to estimate soil humus content. Chapter 11 explores these two conclusions in the context of both aircraft and satellite sensor data.

PART III
Spectral reflectance of vegetation

6
Modelling of vegetation canopy reflectance

Analysis of vegetation by remote sensing and the need for standardised data processing require the development of physiomathematical models of the interaction of radiation with a vegetation canopy (Slater, 1980). The limited application of radiation transfer theory has resulted in increased attention being paid, at least in the USSR, to the use of models based on radiation field theory, and for this reason such work is reviewed in this chapter.

6.1. Approaches to the modelling of reflectance from vegetation canopies in the USSR

A vegetation canopy can be considered to be a randomly heterogeneous medium (Kozoderov, 1982) comprising a totality of elements (leaves, branches, etc.) each of which has a definite geometry and distribution of dielectric properties. For the models of dielectric penetrability of individual elements a wave equation, which describes the interaction of a radiation field with a canopy can be solved by Green's functions. Multiple scattering of radiation by the various components of a canopy can be accounted for in Feinman's diagram method (Bass and Fuks, 1972; Rytov *et al.*, 1978; Isimaru, 1981a) developed in quantum electrodynamics. The necessary information on the canopy's dielectric properties, physiochemical and biometric parameters and their respective distributions can be derived from laboratory and field measurements. The radiation measured by the sensor is effectively a smoothed estimate of the radiation field and, as such, details within this field are lost and modelling is, in some cases, simplified.

To understand how a radiation field interacts with a canopy, many researchers in the USSR have used Dyson's integral equation for a mean radiation field and Boethe-Solpeter's equation for the radiation field correlation function (coherence function) (Kozoderov, 1982). There is a problem in that the second momenta of the radiation field is used, but the Gaussian character of the radiation field is temporarily assumed. However, in the future it may be possible to generalise the theory of Feinman's diagrams for non-Gaussian random radiation fields (Finkelberg, 1967), although this presents some theoretical difficulties.

Developing realistic models to describe the distribution of a canopy's dielectric properties is a problem. Parameterisation of vegetation properties, is often based on measurements of height, thickness, per cent coverage, etc., or a statistical statement of the spatial distribution of phyto-elements, leaf area index, leaf orientation functions and various descriptors of canopy architecture (Ross, 1975), none of which is a unique descriptor of a canopy.

Studies of wave diffraction (Born and Wolf, 1973) on a heterogeneous canopy have been the bases for describing mathematically correlations between remotely sensed data and vegetation parameters. For diffraction from a heterogeneous canopy, the problem is a statistical one and involves finding the probabilistic characteristics of the scattered radiation field (here the first and second momenta of the radiation field) from the known statistical characteristics of the canopy (distributions of dielectric penetrability). A solution to this problem that also accounts for multiple scattering became possible as a result of theoretical developments on wave diffraction by non-fluctuating objects of a simple shape (Kauli, 1979).

6.2. Three approaches to modelling the reflectance from vegetation canopies

There are three approaches to the characterisation of radiation diffraction from randomly heterogeneous canopies (Bass and Fuks, 1972; Rytov et al., 1978; Isimaru, 1981a). These assume either a random cloud of discrete scatters, a random continuous medium or rough surfaces. For each of these approaches, the shape and position of interfaces with different dielectric properties (leaves, branches, etc.) are assumed to be random. The large number of random components associated with the canopy complicates both the perception and analysis of how radiation interacts with it. However, by using the theory of Feinman's diagrams it is possible to assess the influence of a canopy's properties on remotely sensed measurements.

The equations for the interaction of the radiation field with a random cloud of scatterers, using Isimaru's (1981a) notation, are for an average field

$$\langle \Psi^a \rangle = \Phi_i^a + \int U_s^a \langle \Psi^a \rangle \rho(\mathbf{r}_s) \, \mathrm{d}\mathbf{r}_s \qquad (6.1)$$

for the mean-square field module

$$\langle |\Psi^a|^2 \rangle = |\langle \Psi^a \rangle^2| + \int |V_s^a| \langle |\Psi^s|^2 \rangle \rho(\mathbf{r}_s) \, \mathrm{d}\mathbf{r}_s \qquad (6.2)$$

where

$$V_s^a = U_s^a + \int U_t^a U_s^t \rho(\mathbf{r}_s)\, d\mathbf{r}_t \tag{6.3}$$

Ψ^a is the scalar field in the point $\mathbf{r}_a$ of the space not filled with scatterers; Φ_i^a is the wave incident to the point $\mathbf{r}_a$ in the absence of scatterers U_s^a is the wave in the point $\mathbf{r}_a$ scattered by a scatterer located in the point $\mathbf{r}_s$; $\rho(\mathbf{r}_s)$ is the particles' concentration (number of scatterers per unit volume); $\rho(\mathbf{r}_s) = W(\mathbf{r}_s)N$, where N is the total number of scatterers on the assumption that all of the scatterers have similar statistical characteristics; $W(\mathbf{r}_s)\,d\mathbf{r}_s$ is the probability of finding the sth scatterer in the elemental volume $d\mathbf{r}_s$; $\langle |\Psi^a|^2 \rangle$ is the total radiance equal to the sum of coherent $|\langle \Psi^a \rangle|^2$ and non-coherent $\langle |\Psi_f^a| \rangle$ radiance; angle brackets denote averaging of the respective quantities; s is the current index for all of the scatters; the index t marks all the scatterers except for s and the index a characterises a wave at the point $\mathbf{r}_a$.

Two similar equations for random continuous media are (Isimaru, 1981a) Dyson's equation for Green's average function G:

$$\bar{G}(\mathbf{r}, \mathbf{r}_0) = \bar{G}_1(\mathbf{r}, \mathbf{r}_0) + \int \bar{G}_1(\mathbf{r}, \mathbf{r}_1)[Q(\mathbf{r}_1, \mathbf{r}_2) - Q_1(\mathbf{r}_1, \mathbf{r}_2)]\bar{G}(\mathbf{r}_2, \mathbf{r}_0)\, d^3\mathbf{r}_1\, d^3\mathbf{r}_2 \tag{6.4}$$

and Boethe-Solpeter's equation for the coherence function $\Gamma(\mathbf{r}', \mathbf{r}'') = \langle \Psi(\mathbf{r}')\Psi^*(\mathbf{r}'')\rangle$ of the Ψ scalar field

$$2\nabla_\mathbf{R}\nabla_\mathbf{r}\Gamma(\mathbf{R},\mathbf{r}) = k^4 \int [\bar{G}(\mathbf{r}) - \bar{G}^*(\mathbf{r}')][B_\varepsilon(\mathbf{r}') - B_\varepsilon(\mathbf{r} - \mathbf{r}')]\Gamma(\mathbf{R}, \mathbf{r} - \mathbf{r}')\, d^3\mathbf{r}' \tag{6.5}$$

where $\mathbf{r}_0$ characterises the position of a point source; Q is the mass operator nucleus; $\mathbf{R} = \frac{1}{2}(\mathbf{r}' + \mathbf{r}'')$, $\mathbf{r} = \mathbf{r}' - \mathbf{r}''$, $2\nabla_\mathbf{R}\nabla_\mathbf{r} = \Delta' - \Delta''$, where Δ' and Δ'' are Laplace operators for the points $\mathbf{r}'$ and $\mathbf{r}''$, respectively; $\varkappa = 2\pi/\lambda$ is the wavenumber (λ is the wavelength); $\bar{G}$ and $\bar{G}^*$ are Green's mean function and the complex conjugated function is represented as an averaged series from the theory of disturbances; the index 1 for G and Q corresponds to a definite approximation in the solution of the problem (mostly Burre approximation, discussed below); $B_\varepsilon(\mathbf{r}', \mathbf{r}'') = \langle \varepsilon(\mathbf{r}')\varepsilon^*(\mathbf{r}'')\rangle - \langle \varepsilon(\mathbf{r}')\rangle \langle \varepsilon(\mathbf{r}'')\rangle$ is the field of dielectric penetrability of the canopy ε and for a statistically homogeneous canopy, for which Equations 6.4 and 6.5 hold, $B_\varepsilon(\mathbf{r}', \mathbf{r}'') = B_\varepsilon(\mathbf{r}' - \mathbf{r}'')$.

Respective equations for rough surfaces (Bass and Fuks, 1972) may be written as

$$\bar{G}(\mathbf{r}, \mathbf{r}_0) = G_0(\mathbf{r}, \mathbf{r}_0) + \int G_0(\mathbf{r}, \mathbf{r}_1)\, Q(\mathbf{r}_1, \mathbf{r}_2)\, \bar{G}(\mathbf{r}_2, \mathbf{r}_0)\, d^3\mathbf{r}_1\, d^3\mathbf{r}_2 \tag{6.6}$$

$$\Gamma(\mathbf{r}', \mathbf{r}''; \mathbf{r}_0', \mathbf{r}_0'') = \bar{G}(\mathbf{r}', \mathbf{r}_0') \, \bar{G}^*(\mathbf{r}'', \mathbf{r}_0'') + \int \bar{G}(\mathbf{r}', \mathbf{r}_1) \, \bar{G}^*(\mathbf{r}'', \mathbf{r}_2) \, K(\mathbf{r}_1, \mathbf{r}_2; \mathbf{r}_3, \mathbf{r}_4)$$
$$\times \Gamma(\mathbf{r}_3, \mathbf{r}_4; \mathbf{r}_0', \mathbf{r}_0'') \, d^3\mathbf{r}_1 \, d^3\mathbf{r}_2 \, d^3\mathbf{r}_3 \, d^3\mathbf{r}_4 \quad (6.7)$$

where K is the radiance operator nucleus

$$\Gamma(\mathbf{r}', \mathbf{r}''; \mathbf{r}_0', \mathbf{r}_0'') = \langle G(\mathbf{r}', \mathbf{r}_0') \, G^*(\mathbf{r}'', \mathbf{r}_0'') \rangle \quad (6.8)$$

is the Green's function G after allowing for disturbances in boundary conditions caused by the effect of the rough surface on the radiation field.

Thus, three different integral equations describe the same physical process of the interaction between waves and the three types of randomly ordered canopy. Each is related to specific features of disturbance introduced to the radiation field by the canopy.

The phenomenological theory of radiation transfer, which also takes into account multiple scattering, could be considered as an alternative to the theory of multiple scattering. For a random cloud of scatterers the amplitude of scattering $f(\mathbf{e}_s, \mathbf{e}_i)$, defines the scattering field in a far zone as both a spherical wave and the effective constant of propagation of the mean (scattered) field:

$$\tilde{\varkappa} = k + [2\pi \rho f(\mathbf{e}_s, \mathbf{e}_i)/k] \qquad \mathrm{Im}\,\tilde{\varkappa} = \rho \sigma_t / 2$$

where $(\mathbf{e}_s, \mathbf{e}_i)$ are unit vectors in the directions of scattering and incidence and σ_t is the total interaction cross section. In this case Equation 6.3 is written as

$$V_s^a = f(\mathbf{e}_s, \mathbf{e}_i) \exp(i\tilde{k} |\mathbf{r}_a - \mathbf{r}_s|) / |\mathbf{r}_a - \mathbf{r}_s| \quad (6.9)$$

The second eigenvector of the field is related to the beam radiance of the transfer equation as (Isimaru, 1981a, b)

$$\langle \Psi(\mathbf{r}_a) \, \Psi^*(\mathbf{r}_b) \rangle = \Gamma(\mathbf{r}_a, \mathbf{r}_b) = \Gamma(\mathbf{r}_a, \mathbf{r}_b) \approx \int I(\mathbf{r}, \mathbf{s}) \exp(i\tilde{k}_r \mathbf{s} \mathbf{r}_d) \, d\Omega \quad (6.10)$$

where $\mathbf{r} = \tfrac{1}{2}(\mathbf{r}_a + \mathbf{r}_b)$, $\mathbf{r}_d = \mathbf{r}_a - \mathbf{r}_b$, $\tilde{k}_r = \mathrm{Re}(\tilde{k})$; $d\Omega$ is the element of solid angle; $\mathbf{s}$ is the unit vector of mean density of the energy flux in the point $\mathbf{r}$. The radiance $\langle |\Psi(\mathbf{r})|^2 \rangle$ is determined from

$$\langle |\Psi(\mathbf{r})|^2 \rangle = \int I(\mathbf{r}, \mathbf{s}) \, d\Omega \quad (6.11)$$

It follows from Equations 6.9 and 6.10 that

$$|V_s^a|^2 \langle |\Psi^s|^2 \rangle = \int |f(\mathbf{s}, \mathbf{s}')|^2 \frac{\exp(-\rho \sigma_t |\mathbf{r}_a - \mathbf{r}_s|)}{|\mathbf{r}_a - \mathbf{r}_s|^2} I(\mathbf{r}_s, \mathbf{s}) \, d\Omega' \quad (6.12)$$

where $\mathbf{e}_s$ and $\mathbf{e}_i$ unit vectors are substituted by $\mathbf{s}$ and $\mathbf{s}'$, respectively.

Introducing the phase function

$$\gamma(\mathbf{s}, \mathbf{s}') = \frac{4\pi}{\sigma_t} |f(\mathbf{s}, \mathbf{s}')|^2 \quad (6.13)$$

and mean intensity

$$\bar{I}(\mathbf{r}) = \frac{1}{4\pi} \int_{4\pi} I(\mathbf{r}, \mathbf{s}) \, d\Omega \quad (6.14)$$

gives the radiation transfer equation

$$\bar{I}(\mathbf{r_a}) = I_{ri}(\mathbf{r_a}) + \int d\mathbf{r_s} \frac{\exp(-\rho\sigma_t |\mathbf{r_a} - \mathbf{r_s}|)}{4\pi |\mathbf{r_a} - \mathbf{r_s}|^2} \int \frac{\rho\sigma_t}{4\pi} \gamma(\mathbf{s}, \mathbf{s}') \bar{I}(\mathbf{r_s}, \mathbf{s}') \, d\Omega' \quad (6.15)$$

where the coherent radiance $|\langle \Psi^a \rangle|^2$ is reduced in the same way as a reduced incident intensity, characterised by the index i.

For a continuous canopy (Rytov et al., 1978) a Fourier transformation of the coherence function can be attempted

$$\Gamma(\mathbf{R}, \mathbf{r}) = \int \tilde{\Gamma}(\mathbf{R}, \mathbf{x}) \exp(i\mathbf{x}\mathbf{r}) \, d^3x \quad (6.16)$$

with the use of Green's mean function represented as

$$G(\mathbf{r}) = -\frac{\exp(i\tilde{k}r)}{4\pi r} \quad (6.17)$$

where $\tilde{k} = k_1 + ik_2$ ($k_2 \ll k_1$) the spectral difference density can be obtained

$$Z(\mathbf{x}) = \frac{1}{8\pi^3} \int [\bar{G}(\mathbf{r}') - \bar{G}^*(\mathbf{r}')] \exp(-i\mathbf{x}\mathbf{r}) \, d^3r$$

$$= \frac{2k_1 k_2}{4\pi^3 i[(k_1^2 - k_2^2 - x^2)^2 + 4k_1^2 k_2^2]} \quad (6.18)$$

Bearing in mind that $k_2 \ll k_1$, the following representation of the $Z(x)$ function in terms of the delta-function δ can be derived:

$$Z(x) \approx \frac{\delta(x - k_1)}{8\pi^2 i k_1} \quad (6.19)$$

Since

$$\int B_\varepsilon(\mathbf{r}') \exp(i\mathbf{x}\mathbf{r}') \, d^3r' = 8\pi^3 \Phi_\varepsilon(\mathbf{x}) \quad (6.20)$$

where $\Phi_\varepsilon(\mathbf{x})$ is the spectral density (spatial spectrum) of a random homogeneous field, it follows from Equation 6.5 that

$$\mathbf{x}\nabla_R(\mathbf{R},\mathbf{x}) = \frac{\pi k^4}{2k_1} \tilde{\Gamma}(\mathbf{R},\mathbf{x}) \oint \Phi_\varepsilon(k_1\mathbf{n}' - \mathbf{x}) \, d\Omega(\mathbf{n}')$$

$$+ \frac{\pi k^4}{2k_1} \delta(x - k_1) \int d^3x' \tilde{\Gamma}(\mathbf{R},\mathbf{x}') \Phi_\varepsilon(\mathbf{x} - \mathbf{x}') \quad (6.21)$$

where $\mathbf{n}'$ is the unity vector; $x' = x'\mathbf{n}'$, $d^3x' = x'^2 \, dx' \, d\Omega(\mathbf{n}')$.

For a homogeneous and stationary medium, when the dispersive hypersurface in the space (w, k_1) is a core $x^2 = w^2/c^2 = k_1^2$, or $|x| = k_1$ (where w and c are the wave frequency and speed, respectively), the intensity (brightness) $I(\mathbf{R},\mathbf{n})$ is introduced, so that

$$\Gamma(\mathbf{R},\mathbf{x}) = \delta(x - k_1) \, I(\mathbf{R},\mathbf{n}) \frac{1}{k_1^2} \quad (6.22)$$

Again this leads to the radiation transfer equation

$$\mathbf{n}\nabla_R I(\mathbf{R},\mathbf{n}) = -\alpha I(\mathbf{R},\mathbf{n}) + \oint \sigma(\mathbf{n},\mathbf{n}') \, I(\mathbf{R},\mathbf{n}') \, d\Omega(\mathbf{n}') \quad (6.23)$$

where

$$\alpha = \oint \frac{\pi k^4}{2} \Phi_\varepsilon(k_1,\mathbf{n}') \, d\Omega(\mathbf{n}') \quad (6.24)$$

is the reduction coefficient, similar to the total interaction cross section mentioned above and

$$\sigma(\mathbf{n},\mathbf{n}') = \frac{\pi k^4}{2} \Phi_\varepsilon[k_1(\mathbf{n} - \mathbf{n}')] \quad (6.25)$$

is the cross section of scattering from a unit volume of the continuous canopy to a unit solid angle.

It follows from Equations 6.16 and 6.22 that

$$\Gamma(\mathbf{R},\mathbf{r}) = \frac{1}{k_1^2} \int e^{i\mathbf{x}\mathbf{r}\mathbf{n}} \delta(x - k_1) I(\mathbf{R},\mathbf{n}) x^2 \, dx \, d\Omega(\mathbf{n})$$

$$\text{or} \quad \Gamma(\mathbf{R},\mathbf{r}) = \oint I(\mathbf{R},\mathbf{n}) e^{ik_1\mathbf{r}\mathbf{n}} \, d\Omega(\mathbf{n}) \quad (6.26)$$

i.e. the beam radiance is the angular spectrum of the coherence function.

These expressions are valid with the following assumptions: the $\Gamma(\mathbf{R}, \mathbf{r})$ function in the Fourier transformation accounts for only the $\exp(i\varkappa r)$ plane waves for which $|\varkappa| = k_1$; the first terms in the extension of the mass operator nucleus and intensity operator nucleus are the so-called Burre approximation and staircase approximation and the scale of the $\Gamma(\mathbf{R}, \mathbf{r})$ function by the variable $\mathbf{r}$ is small compared with its characteristic scale by the variable $\mathbf{R}$. The last two assumptions hold when

$$k^2 l_\varepsilon^2 \tilde{\sigma}^2 \ll 1$$

where l_ε is the correlation radius of the dielectric penetrability field ε and $\tilde{\sigma}^2$ is the dispersion of this field.

For a random rough surface (Bass and Fuks, 1972) the transfer equation is obtained upon averaging the $\Gamma(\mathbf{r}', \mathbf{r}''; \mathbf{r}_0', \mathbf{r}_0'')$ function over the physical interval L, the size of which satisfies the inequalities

$$l_\varepsilon, \frac{1}{(x_m - x_n)_{\min}} \ll \tilde{L} \ll \frac{1}{(\gamma_n)_{\max}} \tag{6.27}$$

where γ_n^{-1} is the distance at which the mean (coherent) field of the nth wave (with a wavenumber x_n) varies substantially with its transformation to other modes. This averaging can be made for the canopy, the rough boundary of which does not lead to a substantial reconstruction of the rough surface spectrum compared with a smooth surface. In optical wavelengths the sensor generally performs this averaging automatically. Remotely sensed measurements are suitable for the validation of radiation transfer theory, but there are cases when the theory is not suitable for validating remotely sensed measurements (Gazarian, 1969; Barahanenkov, 1973). For instance, it is not valid for backscattering measurements (Rytov et al., 1978).

Here we report an investigation of a canopy that was modelled as having rows with rough top and bottom boundaries in which two-dimensional (cylindrical) roughnesses were considered with rows parallel to the y-axis. In the first approximation such a canopy can be considered as a waveguide, as discussed in Nikolsky (1978).

For this canopy model the radiation field $I(\mathbf{r}', \mathbf{r}_0'') = \Gamma(\mathbf{r}, \mathbf{r}''; \mathbf{r}_0', \mathbf{r}_0'')$, after averaging over the interval L (along the x-axis) with disturbances of the field by a point source located at (x_0, z_0), is

$$I(\mathbf{r}', \mathbf{r}_0') = 2\pi \sum_n \frac{\phi_n^2}{x_n} 2\pi \sum_m \frac{\phi_m^2}{x_m} F_{nm}(x, x_0) \tag{6.28}$$

where x_n and $\phi_n(z)$ are eigenvalues and eigenfunctions of the differential equation

$$\left(\frac{d^2}{dz^2} + k^2 \varepsilon(z) - x_n^2\right) \phi_n(z) = 0 \tag{6.29}$$

F_{nm} satisfies the system of equations obtained on integrating Equation 6.7 in the x-derivative of the $\tilde{L}$ interval

$$F_{nm}(x, x_0) = F^0_{mn}(x - x_0) + \sum_{p,q=\tilde{N}}^{\tilde{N}} \int_{-\infty}^{+\infty} dx_1 F^0_{np}(x - x_1) \tilde{B}(x_p - x_q) F_{qm}(x_1, x_0) \quad (6.30)$$

where F^0_{mn} refers to a smooth waveguide, $\tilde{N}$ is the number of waveguide modes and $\tilde{B}$ is the Fourier transformation of the correlation function S for a rough surface $z = \varsigma(x)$

$$B(x_1, x_2) = \frac{1}{\sigma^2} \langle \varsigma(x_1) \varsigma(x_2) \rangle \quad (6.31)$$

The transfer equation for intensity I (remember that this is usually radiance) is obtained upon differentiating Equation 6.30 with the use of the Burre approximation and the optical theorem (Bass and Fuks, 1972; Isimaru, 1981a)

$$\frac{x_n}{x_n} \nabla_r I(\mathbf{x}_n, \mathbf{r}) = -2\gamma^0(\mathbf{x}_n) I(\mathbf{x}_n, \mathbf{r})$$

$$+ \frac{1}{2\pi} \int_{-\pi}^{\pi} d\alpha \sum_{m=1}^{\tilde{N}} \tilde{W}(\mathbf{x}_n, \mathbf{x}_m)[I(\mathbf{x}_m, \mathbf{r}) - I(\mathbf{x}_n, \mathbf{r})] \quad (6.32)$$

where γ^0 is the wave extinction characteristic considering both variations in the mean field and energy leakage caused by absorption by the waveguide's walls and penetration of the radiation field through the surfaces confining the waveguide, α is the angle between the x_n and x_m vectors

$$\gamma_n = \frac{1}{2} \sum_{m=-\tilde{N}}^{\tilde{N}} \tilde{W}_{nm}$$

and $\tilde{W}_{nm}$ is the probability of energy transfer from the nth mode to the mth mode at a single scattering.

These calculations were confined to the case of a wide multimode waveguide with large-scale roughness and the absence of shading. The calculations included the following observational data: wavelength, object height, surface shape, angles of the slopes and radii of the roughness features. These calculations represent a relatively simplified situation compared with that in nature. This is necessary because of the difficulties in describing the roughness of the medium and in exciting a waveguide by radiation from every part of the sky.

Figure 6.1 Relationship between the brightness coefficient ($\lambda = 0.675 \mu m$) of a vegetation canopy and (a) the height of plants in that canopy, (b) the density of plants in that canopy and (c) the proportion of that canopy that does not contain plants. Where I is the intensity of upwelling radiation from a canopy, I_r is the intensity of upwelling radiation from a reference target, I_0 is the intensity of downwelling radiation, D is directional radiation and H is global radiation (observed relationship only). In each case (1) shows the theoretical and (2) the observed relationship.

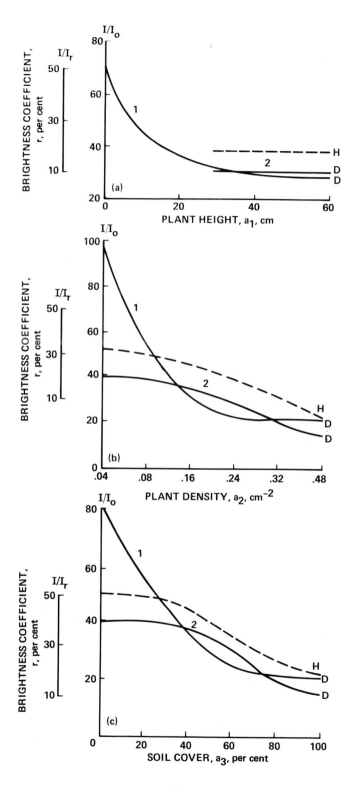

6.3. A field evaluation of reflectance models for vegetation canopies

In summer 1981 a field study was performed at the Institute for Agricultural Meteorology (Obninsk) to investigate some of the relationships between vegetation and remotely sensed data (Kondratyev and Fedchenko, 1980a). In the experiment it was possible to vary the density, height and cover of plants. The remote sensing measurements were made with a lens photometer.

Results of measurements and calculations from theory, for a wavelength of 0.675 μm, are given in Figure 6.1. These results were presented as

$$I = f(a_1, a_2, a_3) \qquad (6.33)$$

where I is the measured intensity (brightness) from the surface (Table 1.1), a_1 is the height of the plants, a_2 is the plants' density and a_3 is the soil cover (1 − vegetation cover). The averages of the parameters a_1, a_2 and a_3 were, respectively, 30 cm, 0.2 cm^{-2} and 50 per cent in Figure 6.1. The trends of the reflectance curves in Figure 6.1 obtained both from measurement and calculated from theory were very similar. A comparison of the curves in Figure 6.1c shows that to consider a canopy as a set of rough random surfaces holds where there is 50–80 per cent soil cover (Figure 6.1c). Below 50 per cent soil cover this system should, apparently, be modelled by a random cloud of discrete volume scatterers. At 80 per cent or more soil cover the canopy could be considered as a continuous medium. Note that it was assumed in the calculations that $\varepsilon(z)$ was a linear function. In fact, $\varepsilon(z)$ is a random function which makes it necessary to solve stochastic equations such as Equation 6.29, the solutions to which are only known for certain cases (Kliatskin, 1980).

The theoretical aspect of vegetation canopy modelling could be improved by accounting for, amongst other variables, the vector character of electromagnetic waves with changing polarisation at multiple scattering (Rytov et al., 1978; Talmage and Curran, 1986), by considering shading, modulation of heterogeneities of the medium by wind roughness (Isimaru, 1981a), as well as non-monochromacity of the incident waves (Akhmanov et al., 1981).

6.4. Concluding comments on modelling vegetation canopy reflectance

Radiation field theory has provided researchers in the USSR with an increased understanding of how radiation interacts with vegetation canopies. This understanding is the basis of several experimental designs reported in the following chapters. In these experiments remotely sensed data are used to estimate the chlorophyll content, state and weed component of vegetation canopies.

7
Remote sensing of crop chlorophyll

The chlorophyll concentration of leaves and its estimation are important topics in plant physiology and ecosystem ecology. Many reliable techniques exist to determine chlorophyll concentration; however, they are relatively laborious and time-consuming, and they destroy the sample. Consequently, simpler non-destructive approaches for estimating chlorophyll concentrations have been sought. For instance, Benedict and Swidler (1961) found that the chlorophyll concentration of soybean leaves was highly correlated with absorption coefficients at a wavelength of 0.625 μm, and that the absorption coefficient at this wavelength could be used to estimate chlorophyll concentration. This, and similar results, enabled the development of indirect, optical, nondestructive and now-standard techniques to estimate chlorophyll concentration (Horwitz, 1970). However, such techniques are effectively limited to laboratory conditions, because it is extremely difficult to measure absorption coefficients in the field.

The reflection coefficient may be a more representative physical quantity for indirectly estimating plant chlorophyll concentration in both laboratory and field conditions (Benedict and Swidler, 1961; Curran and Milton, 1983). However, experiments have shown that the reflectance coefficient measured at 0.625 μm is not always a reliable variable for this purpose. Furthermore, inspection of spectral reflectance curves for different plants has shown that it is unlikely that measurements acquired at any one wavelength or at any one point on the spectral reflectance curve will provide a reliable estimation of chlorophyll concentration.

It is known that even small variations in the chlorophyll concentration can modulate the spectral-reflectance curve, particularly in green wavelengths. Furthermore, total reflectance typically declines with increasing chlorophyll concentration and there does not appear to be an asymptote, i.e. a threshold concentration above which the spectral reflection coefficient remains constant (Brandt and Tageeva, 1967). Because a change in chlorophyll concentration introduces a change in the shape of the spectral reflectance curve, it may be possible to estimate the chlorophyll concentration using data on the spectral reflectance of leaves and canopies.

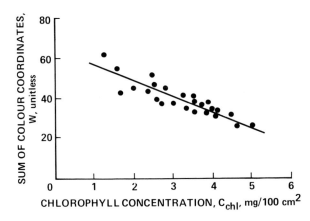

Figure 7.1 The relationship between the sum of the colour coordinates and the chlorophyll concentration of potato leaves.

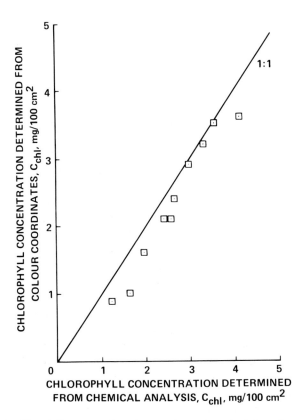

Figure 7.2 The chlorophyll concentration of potato leaves determined form the sum of the colour coordinates and by conventional chemical analysis.

Results of a preliminary experiment reported here indicate that it was indeed possible to use the spectral reflectance of leaves, via quantitative colorimetry, to estimate chlorophyll concentration (Kondratyev et al., 1982a). The spectral reflectance of 1–2 cm diameter circular samples of potato leaves were measured with an SF-18 spectrophotometer. The chlorophyll concentration of each sample was then measured by the conventional chemical analysis described by Baslavskaya and Trubetskova (1964).

Colour coordinates were calculated from the spectral reflectance data using Equation 2.9 and the sum of all three coordinates was then determined

$$W = X + Y + Z$$

There was strong linear correlation ($r = 0.91$) between W and the measured chlorophyll concentration (Figure 7.1). Furthermore, a comparison between the estimated and measured chlorophyll concentrations showed reasonable agreement (Figure 7.2).

7.1 The influence of species-specific leaf structure on the relationship between colour coordinates and chlorophyll concentration

The experiment above showed that the chlorophyll concentration of green leaves can be estimated from colour coordinates. However, this used the entire spectral-reflectance curve, which is influenced by a variety of plant elements. Consequently, it is not always a reliable technique and may not be able to distinguish the relative magnitude of chlorophyll concentration for a group of samples in which these other plant elements vary. For instance, Figure 7.3 illustrates the spectral-reflectance curves for leaves with different chlorophyll concentrations, and demonstrates that the negative relationship between chlorophyll concentrations and reflectance is not an absolute one, especially in visible wavelengths (Wooley, 1971). Estimation of chlorophyll concentration using the colour coordinates derived from these curves is likely to be inaccurate, essentially because the chlorophyll concentration is not the only factor determining leaf reflectance.

To investigate the other factors influencing leaf reflectance, a laboratory experiment was performed. The spectral reflectance of buckwheat, potato, sorgo and red clover leaves was measured over the $0.40–0.75$ μm range with an SF-18 spectrophotometer. These mesurements were repeated after the leaves had been placed in alcohol to remove the chlorophyll.

The effect of chlorophyll removal on the reflectance of buckwheat and barley was marked (Figure 7.4). The green leaves displayed the typical curve for vegetation reflectance with reflection minima in blue and red and maxima in green and near-infrared wavelengths. In the absence of chlorophyll the reflectance of the leaves was less selective. The maximum in green and minimum in blue wavelengths were no longer apparent; the minimum in red wavelengths was less apparent (the absorption feature in the treated leaf was probably due to incomplete chlorophyll extraction) and the maximum in near-infrared wavelength was considerably reduced, probably as a result of structural damage to the leaf.

The SRC was not the same for the four species of chlorophyll-free leaves (Figure 7.5). It is therefore likely that the correlation between leaf spectral reflectances and chlorophyll concentration is species-specific.

An experiment to estimate the errors introduced by species-specific leaf components into the estimation of chlorophyll concentration was performed. Leaves with different chlorophyll concentrations were chopped; their spectral reflectance was measured with an SF-18 spectrophotometer and the chlorophyll

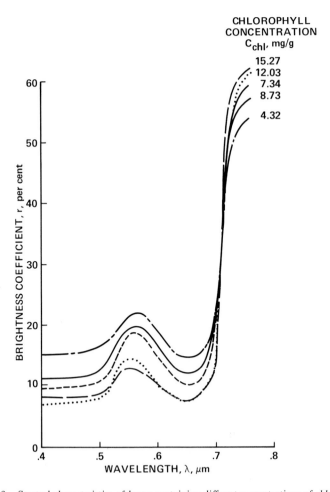

Figure 7.3 Spectral characteristics of leaves containing different concentrations of chlorophyll.

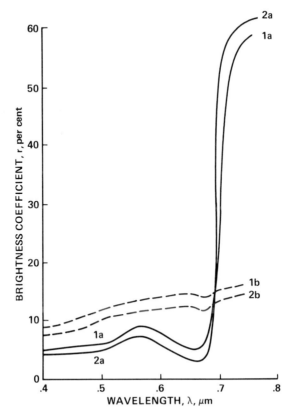

Figure 7.4 Spectral characteristics of the leaves of (1) buckwheat and (2) barley, (a) before and (b) after chlorophyll removal.

concentration was measured chemically. The sum of the colour coordinates, W, derived from the spectral-reflectance curves (Kondratyev *et al.*, 1982a,b,c), was correlated with the chemical measurements of chlorophyll concentration (Figure 7.6).

The spectral reflectance of chopped leaves from the other crops was then measured and their colour coordinates were determined and summed. These data were then used to estimate the chlorophyll concentration using the calibration relationship of Figure 7.6, which had been derived for buckwheat. The results of this are given in Figure 7.7. The next three stages of the experiment were: (1) to extract the chlorophyll by placing the leaves in alcohol; (2) to measure the amount of extracted chlorophyll (Baslavskaya and Trubetskova, 1964); and (3) to measure the spectral reflectance and derive the sum of the colour coordinates for the chlorophyll-free leaves.

Figure 7.7 shows that the spectral reflectance of leaves, expressed here as colour coordinates, is dependent not only on the chlorophyll concentration, but also on the spectral reflectance of the species-specific leaf structure. Therefore,

Spectral reflectance of vegetation

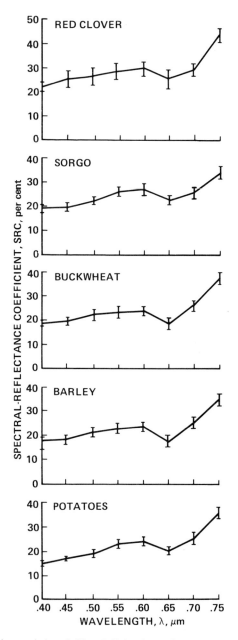

Figure 7.5 Spectral characteristics of chlorophyll-free leaves from five crops with their 70 per cent confidence limits (n = 5) plotted as error bars.

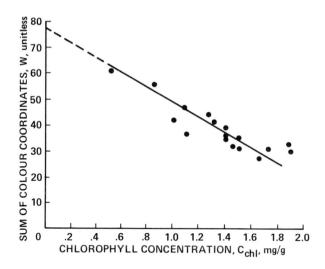

Figure 7.6 The relationship between the sum of the colour coordinates and the chlorophyll concentration of buckwheat leaves.

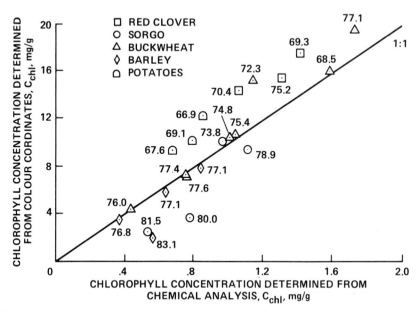

Figure 7.7 The chlorophyll concentration of leaves from five species of crop, determined from the sum of the colour coordinates and by conventional chemical analysis. The value associated with each point is the sum of the colour coordinates for the chlorophyll-free leaves.

the accuracy with which chlorophyll concentration can be estimated from colour coordinates is partially dependent on the SRC of this species-specific leaf structure.

The W, at zero chlorophyll concentration, can be estimated at 76·5 by extrapolating the relationship, illustrated in Figure 7.6, to the ordinate axis. Figure 7.7 shows that the chlorophyll concentrations estimated from the colour coordinates and measured in the laboratory are very similar when the sum of the colour coordinates for the chlorophyll-free leaves was close to 76·5; otherwise they differed markedly.

Consequently, the practical application of the colour-coordinate technique for estimation of chlorophyll concentrations should ideally consist of five stages. First, the spectral reflectance of the leaf should be measured and the sum of its colour coordinates derived from its spectral-reflectance curve; secondly the chlorophyll should be extracted from the leaf; thirdly, the spectral reflectance of the chlorophyll-free leaf should be measured and the sum of the colour coordinates at zero chlorophyll concentration (W_0) calculated; fourthly the relationship between the sum of the colour coordinates for the green leaf and the chlorophyll concentration should be modified to allow for species-specific variability in W_0; and fifthly, the leaf chlorophyll concentration should be estimated from the sum of the colour coordinates of the green leaf and the relationship derived in the fourth stage (Kondratyev et al., 1982a,b,c).

7.2. The relationship between colour coordinates measured in the field, chlorophyll concentration and crop yield

In general crop yield is determined by all the photosynthetically active aboveground organs of a plant (leaves, stalks, spike, etc). Thus, estimates of products should be based not on just the leaf area (Nichiporovich, 1955), but on a chlorophyll index (Tarchevsky, 1977). This is typically expressed in terms of grams of chlorophyll per square meter (or kilograms of chlorophyll per hectare) of crops and averaged over a time period, which has typically been 7–10 days in the USSR. The most representative characteristic that can be used for indirect and non-destructive estimation of the chlorophyll concentration of a plant is likely to be the SBC. The experiments described previously showed that spectral reflectance could be used to estimate the chlorophyll concentration in the laboratory, here the aim was to use remotely sensed spectral reflectance to estimate the chlorophyll concentration and thence yield, in the field.

At 20 points in a barley field a sample of five to seven plants was obtained for analysis in the laboratory. The spectral reflectance of these samples was measured with an SF-18 spectrophotometer and the chlorophyll concentration was measured chemically, as in previous experiments. The relationship between the sum of the colour coordinates (derived from the spectral-reflectance curves) and the measured chlorophyll concentration is shown in Figure 7.8a. Once this laboratory calibration relationship had been quantified, studies to establish the

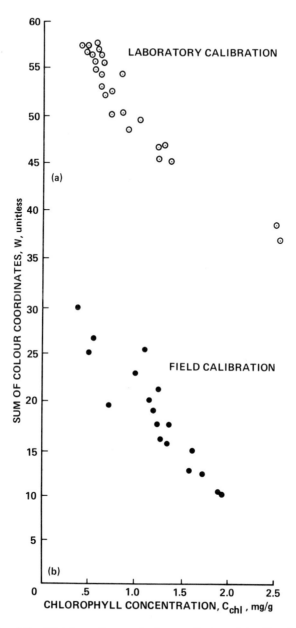

Figure 7.8 The relationship between the sum of colour coordinates and the chlorophyll concentration, (a) is for a laboratory study in which the chlorophyll concentration was measured by conventional chemical analysis and (b) is for a field study in which the chlorophyll concentration was estimated using the relationship shown in scatterplot (a). Note that the sum of the colour coordinate will always be lower in the field environment, as the object of study is a canopy rather than a leaf and other environmental phenomena modulate the relationship.

relationships between field measures of colour coordinates, chlorophyll concentration and crop yield were undertaken.

Forty 1 m × 1 m sample sites in barley crops with different canopy densities but all in the earing phase were identified. At each site the SBC was measured vertically with a lens photometer in eight spectral intervals centred at $0\cdot40$, $0\cdot45$, $0\cdot50$, $0\cdot55$, $0\cdot60$, $0\cdot65$, $0\cdot70$ and $0\cdot75$ μm. To exclude the soil background a screen was set on the photometer's objective, which effectively masked out the near-nadir angles. All of the measurements were made within 1 h, at a time of clear skies and at a solar elevation of approximately $50°$.

The aim was to estimate non-destructively the chlorophyll concentration at these sites and, when the barley ripened, harvest the plants to establish a correlation between the earlier estimate of chlorophyll concentration and crop yield. As with previous investigations, the calibration curve for estimating the chlorophyll concentration from the colour coordinates had to be established in the field and not in the laboratory, because the two are usually dissimilar (Section 4.4).

The field calibration was produced using data from 20 sites. The sites were all in the same field and they spanned the entire range of chlorophyll concentrations expected in this region. After the spectral reflectance of each site had been measured, the plants were cut and analysed in the laboratory. Only the upper two thirds of the plants were analysed, because it is this section that influences the majority of the radiation measured by the photometer. These samples were chopped and mixed and three to five small samples were obtained from each for measuring spectral reflectance. If necessary, the samples were mixed again to ensure that the spectral reflectance for the samples obtained from any one sample site did not differ by more than 10 per cent. The resultant spectral-reflectance curves were used to calculate colour coordinates, the sums of which, in conjuction with Figure 7.8a, were used to estimate the chlorophyll concentration for all 20 samples. The colour coordinates and their sums were then calculated for these samples, using the field-measurement data. The resultant scatter plot (Figure 7.8b) shows the relationship between the sum of the field-derived colour coordinates and the chlorophyll concentration.

This procedure of laboratory and field measurements of the 20 control sites made it possible to unify the laboratory and field systems for measuring spectral reflectance and to use, thereby, the calibration curve to estimate the chlorophyll concentration non-destructively. The next stage was to identify the relationship between the chlorophyll concentration and the crop yield. The barley from each of the control sites was allowed to ripen completely before being cut and threshed. The relationship between the chlorophyll concentration of the crop at the earing phase and the crop yield is shown in Figure 7.9. Note that the correlation between the chlorophyll concentration and the crop yield was low, as yield was probably more a function of other factors (e.g. soil moisture, climate, etc.) than of the chlorophyll concentration. Recent studies in the USSR have confirmed that remotely sensed chlorophyll concentration is an unreliable estimator of crop yield.

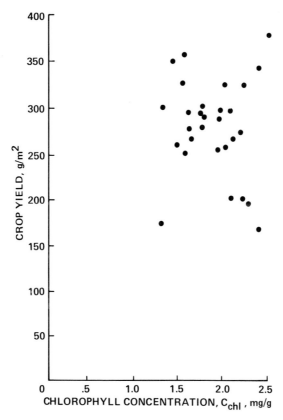

Figure 7.9 The relationship between remotely sensed chlorophyll concentration and crop yield for barley.

7.3. Concluding comments on the remote sensing of crop chlorophyll

Chlorophyll and water account for the major absorption features in the reflectance spectra of vegetation. Therefore, a high correlation between the chlorophyll concentration and the reflectance would be expected, at least for similar canopies. However, the accuracy with which remotely sensed reflectance could then be used to estimate the chlorophyll concentration and thereby infer other variables (e.g. yield) is confounded by a wide range of plant physiological and environmental factors, as this chapter has shown.

Recent evidence points to the value of using spectral shifts, in the $0 \cdot 68$ μm wavelength region for the estimation of chlorophyll concentration (Horler *et al.*, 1983; Rock *et al.*, 1988). Such developments have yet to be reported in the Soviet literature.

8
Remote sensing of crop state

Remote sensing of crop state (or condition) has been reported widely in English-language publications since the 1960s. Most of such work has used remotely sensed reflectance to characterise the state of a wide range of crops, whereas in the USSR attention has been focused on the use of remotely sensed colour to characterise the state of winter crops.

8.1. Remote sensing of winter crop state in spring

A major cause of low crop yield in the USSR is the loss of some winter crops after they have overwintered. It is therefore of considerable agricultural value to develop techniques for assessing the state of winter crops in spring. Presently, in the USSR such procedures are based on field and aerovisual estimates of the area of senesced and damaged crops using a standard five-point scale (Table 8.1; Ogorodnikov, 1971).

These estimates are subjective, and the uneven scale hinders quantitative extrapolation. However, experienced observers can provide estimates that assist yield forecasting. The critical feature in the success (or failure) of these visual assessments is the choice of a colour to indicate crop state. Crops can display a large variety of colours and hues, so large improvements could be made to the method if a clear-cut, quantitative nomenclature of coloration of senesced and unhealthy crops was used. This can be achieved from colorimetry techniques discussed previously.

Although such techniques are able to give a precise colour determination for small objects, such as a leaf, they were not well suited to determining colour for large objects such as fields or forests. Therefore, the colorimetry technique has been refined in the USSR to make it applicable to unevenly damaged agricultural fields. To determine the colour of an object quantitatively, it is necessary to calculate colour and colority coordinates from which, using the colour triangle, a colour hue and purity is found.

To estimate the proportion of unhealthy and healthy crops quantitatively, it is

Table 8.1 *A point scale used by observers in aircraft to estimate state.*

Proportion of brown-to-green or bare-to-green vegetation in the field	Damage to crops	Crop state (points)
0	None	5
<1/5	Some	4
1/3	Considerable	3
4/5	Heavy	2
1	Very heavy	1

not necessary to know their actual coloration (e.g. greyish brown, green, etc.), but it is important to determine a quantitative criterion to express the colour of crops, and this exists in Equation 2.9. Measuring the SBC of an object for input to the equation is not difficult in either the field or the laboratory. The incident spectral radiance can also be determined relatively easily and standard radiation sources (e.g. the D_{65}) have been used to allow precise colour calculations in the laboratory (Chapter 2; Dzhad and Vyshetski, 1978).

To calibrate remotely sensed data in terms of vegetation state, measurements are required for healthy canopies, senesced canopies and bare soil. Field measurements of the SBC were made for 97 sites: 32 were healthy vegetation canopies, 44 were unhealthy vegetation canopies and 21 were bare soils (Kondratyev and Fedchenko, 1982 c,d,g). The spectral reflectance curves obtained from these field-sampled sites were considerably different (Figure 8.1). As expected, the curve for the healthy canopies displayed a maximum in the green and minima in the blue and red spectral regions, and the unhealthy, senesced, canopies and soil exhibited a monotonic increase in reflectance with wavelength (Gausman and Allen, 1973).

The colour coordinates and their variability, illustrated in Figure 8.2, indicate that these three targets are separable from one another. They can thus serve as the basis for determining areas under unhealthy and healthy canopies as

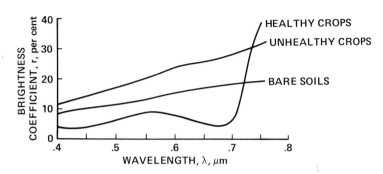

Figure 8.1 *Spectral characteristics of healthy crops, unhealthy crops and bare soils from ground-based measurements.*

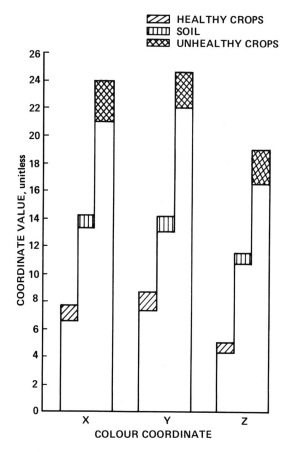

Figure 8.2 Colour coordinates for plants and soils, calculated from their spectral-reflectance curves from ground-based measurements and their 70 per cent confidence limits.

well as areas where the vegetation cover is low and the soil background influences reflectance.

These field-derived calibration relationships were then applied to low-altitude remotely sensed data. Measurements of the SBC (to derive spectral reflectance curves of unhealthy and senescing crops) were obtained in the Kaluga region from a helicopter. Although the movement of the helicopter's rotor blades is known to affect the illumination conditions, it has been shown that these rotor-induced errors in the SBC are negligible and are usually less than 5 per cent (Kondratyev and Fedchenko, 1981b).

All of the measurements were made from an altitude of 120–150 m using a lens photometer (Fedchenko and Kondratyev, 1981) suspended approximately 1 m below the pilot's cabin. Forty-seven fields were sampled and all of the ground- and helicopter-based measurements were recorded under a clear sky at solar elevations above $40°$. The spectral reflectance curves derived from measurements acquired onboard the helicopter were used to calculate the colour coordinates (Equations 2.9) for the soil–crop (unhealthy and healthy) spectral mixtures (Adams et al., 1986). To determine the area under unhealthy and healthy crops, the two-equation spectral-mixture model must be solved.

$$X_{s+c} = S_1 X_{dc} + S_2 X_{hc} + S_3 X_s$$
$$Y_{s+c} = S_1 Y_{dc} + S_2 Y_{hc} + S_3 Y_s \qquad (8.1)$$

Here X_{s+c} and Y_{s+c} are the colour coordinates for the soil–crop (unhealthy and healthy) spectral mixtures; X_{dc} and Y_{dc} are those for unhealthy crops; X_{hc} and Y_{hc} are those for healthy crops; X_s and Y_s are those for soils; and S_1, S_2 and S_3 are the areal proportions of unhealthy and healthy crops and soil, respectively.

As $S_1 + S_2 + S_3 = 1$, a transformation can be made so that Equation 8.1 can be rewritten as

$$X_{s+c} - X_s = S_1(X_{dc} - X_s) + S_2(X_{hc} - X_s)$$
$$Y_{s+c} - Y_s = S_1(Y_{dc} - Y_s) + S_2(Y_{hc} - Y_s) \qquad (8.2)$$

Table 8.2 Calculated cover of winter crops with different degrees of damage in the spring.

Field number	Cover of unhealthy crops (per cent)	Cover of healthy crops (per cent)	Field number	Cover of unhealthy crops (per cent)	Cover of healthy crops (per cent)
1	19	65	25	8	71
2	17	66	26	8	81
3	4	84	27	22	36
4	56	15	28	13	67
5	25	39	29	16	66
6	32	37	30	12	73
7	22	44	31	27	31
8	6	80	32	22	38
9	25	31	33	32	40
10	30	41	34	16	59
11	61	12	35	7	81
12	9	79	36	26	39
13	49	17	37	27	41
14	26	35	38	6	76
15	27	37	39	17	46
16	54	21	40	11	70
17	11	74	41	12	66
18	22	39	42	16	59
19	12	77	43	22	39
20	21	29	44	30	36
21	28	35	45	27	60
22	32	30	46	29	51
23	21	35	47	7	46
24	22	36	48	16	46

Substituting into Equation 8.2 the mean values of the X and Y coordinates taken from Figure 8.2, it is possible to determine the areas under healthy and unhealthy crops and thus to obtain a quantitative estimate of the state of winter crops in a field.

This technique was applied to fields in the Kaluga region, where some of the sampled fields were severely damaged (Table 8.2). This provided data that could be used for agricultural management decisions by focusing attention where it was most needed. For instance, fields 4, 11, 13 and 16 had been ruined, and fields 6, 10, 21, 22, 31 and 44 required urgent attention to improve the crops' condition. Additionally, those fields which were in a satisfactory state of health could also be identified. The technique therefore allows near real-time information on the state of crops and the areal proportions of unhealthy and healthy crops to be assessed. Furthermore, because it is derived from remotely sensed data, this information could be obtained from field to national and even international scales.

In another experiment in the central USSR, a high degree of agreement between the areas of unhealthy crops derived by traditional field analysis and from the colour coordinates was observed for winter crops (Figure 8.3). In Figure 8.3 a value of 100 per cent corresponds to a healthy crop with no soil visible through the vegetation canopy, and values less than 100 per cent show the relative proportion of unhealthy crops at a site. However, the technique does

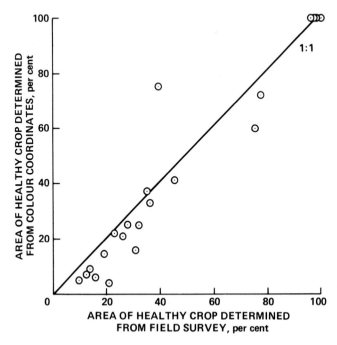

Figure 8.3 Areas under unhealthy crops, determined from the sum of the colour coordinates and from field survey.

require some refinement before it can be applied operationally in the USSR. For example, there are sometimes substantial differences between the areas of unhealthy crops determined by the traditional field method and that from remotely sensed data. This is, in part, due to the subjectivity of the estimates derived in the field and the assumptions made in the remote sensing technique. Nevertheless, the use of an objective quantitative characteristic of crop coloration opens up possibilities for assessing the size of areas under unhealthy and damaged winter crops.

8.2. Remote sensing of winter crop state in autumn

The areas of winter crops that have been damaged in autumn is needed to forecast the wintering potential of crops (Moiseichik, 1978). At present this is assessed visually in the USSR using a subjective indicator such as the evenness of the canopy (similar in concept to Table 8.1). This 'evenness' is, however,

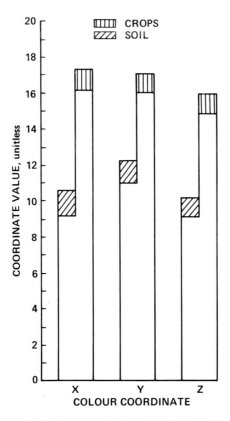

Figure 8.4 *The colour coordinates for crops and soils, calculated from their spectral reflectance curves from ground-based measurements and their 70 per cent confidence limits.*

quantified in terms of the percentage soil cover present, which could be derived from measurements of spectral reflectance. A more objective approach would be to measure the SBC of the soil–crop system with a nadir-viewing airborne photometer. Then, from the known colour coordinates for crops and soils (calculated from *in situ* SBC measurements), the percentage of crop cover could be determined.

Such a procedure was adopted for helicopter-based SBC measurements. At 26 vegetated sample sites and 15 bare-soil sites, ground-based measurement of SBC were acquired. These were used, with Equation 2.9, to calculate the *XYZ* colour coordinates (Figure 8.4).

Measurements of SBC for 24 fields under winter crops and eight bare-soil fields were also made from a helicopter, and the resultant reflectance curves were used in the calculation of colour coordinates. The crop-covered area, S_1, was derived from

$$W_{s-c} = S_1 W_c + S_2 W_s \qquad (8.3)$$

where W_{s-c} is the sum of the colour coordinates for the soil-crop mixtures; W_c and W_s are, respectively, the sums of the colour coordinates for crops and soils; S_1 and S_2 are the proportions of the areal coverage of crops and bare soil, respectively, and $S_1 + S_2 = 1$.

By measuring the SBC for the soil–crop mixtures, deriving the sum of the colour coordinates W_{s-c} and substituting them into Equation 8.3, the W_c and W_s parameters (i.e. the areal coverage of crops and, thereby, bare soils) can be determined (Table 8.3). The areas covered by soils and crops calculated by the traditional field survey and colour-coordinate techniques agree to a large extent (Figure 8.5). However, the latter technique allows a quantitative estimate of crop colour to be used as an indicator of their state.

Absolute statements of field quality cannot be made from the results of

Table 8.3 Calculated percentage cover of winter crops in the autumn.

Field number	Cover of crops	Field number	Cover of crops
1	65	13	42
2	78	14	37
3	82	15	18
4	76	16	21
5	70	17	15
6	13	18	10
7	18	19	10
8	21	20	18
9	15	21	23
10	7	22	36
11	32	23	35
12	35	24	42

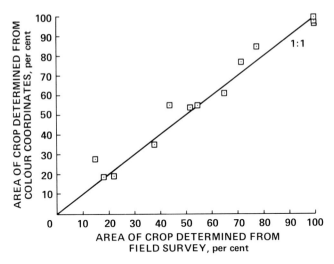

Figure 8.5 Comparison of areas under crops determined from the sum of the colour coordinates and from field survey.

calculations of the area under crops (Table 8.3). To achieve this, quantitative indices that characterise the standards which strictly define what is meant by the state of winter crops in autumn are required for the USSR (Moiseichik, 1978). However, the information on cropped area provides useful information for agricultural management practices.

8.3. Concluding comments on the remote sensing of crop state

The estimation of crop state (or condition) is an important application of remote sensing (Myers, 1983; Ryerson and Curran, 1990). This chapter has identified a combination of spectral mixture modelling and colorimetry for this purpose. This combination forms the basis of much Soviet work on the remote sensing of vegetation (e.g. Chapter 9) which as far as the authors are aware, has yet to be reported in the English-language literature.

9
Remote sensing of crop weeds

Weeds are very important in agriculture as they can significantly reduce crop yield and quality. Dawson and Holstun (1970) identified the major reasons for these reductions.

(1) Inhibition of crop growth as a result of competition with the weeds for water, minerals, light and probably carbon dioxide.
(2) Weeds can give rise to poor conditions for harvesting; e.g. uneven crop growth (this also increases the cost of harvesting).
(3) Attempts to destroy or control weeds may damage the crop.
(4) The presence of weeds in the harvested crop reduces the crop quality.
(5) Weeds act as intermediate hosts for a variety of pests and diseases.

Yield losses in the USSR as a result of weeds can be enormous, with losses in excess of 50 per cent widely reported (e.g. Mani et al., 1968). The damage caused by weeds is dependent on their concentration per unit area (Fisiunov, 1973). For instance, in the USSR a 28–30 per cent reduction in the yield of winter wheat has been associated with 11 sprouts m^{-2} of the common weed *Acroptilon repens*. Increasing the density of *Acroptilon* to 26 m^{-2} reduced the yield by 48–50 per cent and at 60–70 m^{-2} the yield was reduced by 70–75 per cent (Fisiunov, 1973).

9.1. Classification of weeds

In the USSR weeds are generally classified, by the length of their life, into 'one-year' and 'multi-year' weeds. The one-year weeds propagate by seeds and live for usually 1 and not more than 2 years. This group can be further subdivided into three biological groups: spring weeds, wintering weeds and winter weeds. The multi-year weeds can generally be divided into two groups: those that propagate by seeds and those that propagate both vegetatively and by seeds.

Weeds can also be classified into two classes on the basis of feeding characteristics: parasitic and non-parasitic. Parasitic weeds adhere by suction to crops and feed off of them. Weeds which are semiparasitic may be able to live independently of the crop, but they can feed off of it. These are generally bonded to the crop at root level. The non-parasitic weeds do not live off the crop, but synthesise their own organic matter by photosynthesis. These are typically very fertile, as seeds they retain the ability to germinate for a long time and as weeds they are often able to out-compete the crop for minerals, water and light.

9.2. Weed control

Several approaches can be taken to control or destroy weeds. These include a combination of preventive, mechanical (agrotechnical) and chemical measures as well as the use of fire (Fryer and Evans, 1968; van den Bosch and Messenger, 1973; Briggs and Courtney, 1985). There are two broad groups of preventive measures. First, there are techniques which attempt to stop the spread or establishment of weed seeds. These include the purification of seed stock, the removal of seeds from manure and the killing of weeds after crop harvesting and on uncultivated soils. The second set of approaches is aimed at establishing the most favourable conditions for crop as opposed to weed growth and development, including such factors as the correct rotation and careful timing of agricultural practices (Tatarinova et al., 1980).

Agrotechnical methods include modified ways of cultivating the soil (stubble removal, discing, harrowing, ploughing, etc.), crop rotation and careful timing of agricultural operations. The choice of technique or combination of techniques to use depends largely on the characteristics of the crop, climate and soil. However, agrotechnical methods must be carefully applied if they are to be of value, as the result of their application is very time-dependent. Many operations should only be performed, for instance, before sowing or after harvesting, and then only when the soil is at a suitable moisture content (Briggs and Courtney, 1985).

Chemical methods of weed control are used in the USSR when agrotechnical measures are inappropriate. Herbicides are often applied to the crop using an airborne spray early in the crops' growth cycle (if, unlike the weeds, the crop can endure the herbicide). However, this is a generalisation, and the nature of the herbicide, timing and quantity of its application and characteristics of the soil and crop determine how and when chemical control is performed.

Finally, fire is often used to kill weeds, either after harvesting (stubble burning) or in bare fields before sowing. However, this is increasingly subject to legal and environmental controls, especially near major Soviet cities.

9.3. Conventional techniques of recording crop weediness

Application of any of the above techniques is most efficient when the respective agricultural organisations are supplied with accurate and prompt information, preferably in map form, on crop weediness. At present, in the USSR maps of crop weediness are produced by either visual (qualitative) or quantitative (by weight) estimation of weediness. The former can be further subdivided into visual–numerical, visual–projective and visual–combined methods. All of these approaches can be traced back to the visual 'Maltsev scale' that has been used since the early 20th century (Maltsev, 1933). In this approach the degree of weediness is usually estimated using a four-point scale:

(1) low—few weeds, 1 or 2 plants per 100 m^2;
(2) moderate—the weeds are not numerous and usually merge with the crops;
(3) high—weeds are numerous but do not choke the crops;
(4) very high—weeds choke the crops.

This approach is applied to measurements made every 50–100 m along a transect positioned diagonally across a field. The average weediness for each kind of weed can be determined for each field, and this can then be used to produce a map of weediness. However, for tilled crops this approach is difficult to apply, so another weediness scale has gained popularity (Fisiunov and Matiukha, 1972). This classifies the degrees of weediness by the number of weeds per square meter or the number of weed seeds per hectare (Table 9.1).

An elaboration of the earlier Maltsev scale was proposed by Markov (1970). This scale also has four levels of weediness.

(1) Low—few weeds, barely visible among the crops (not more than five sprouts of multi-year weeds or 25 one-year weeds per 100 m^2).
(2) Moderate—weeds clearly visible but, however measured (e.g. number of sprouts, mass, percentage coverage), they are dominated by the crops (5–10 sprouts of multi-year weeds or 25–30 one-year weeds per 100 m^2).
(3) High—weeds are abundant but, however measured, are still dominated by the crops (more than 10 sprouts of multi-year weeds or 50 one-year weeds per 100 m^2).
(4) Very high—weeds dominate the crops.

Further modifications to these scales have been made but have not been widely

Table 9.1 The Fisinov and Matiukha (1972) scale of crop weediness.

Number of weeds per m^2			
One-year weeds	Multi-year weeds	Weed seeds in arable soil layer (million ha^{-1})	Degree of weediness
<10	<1	<10	Low
10–15	1–5	10–50	Moderate
>50	>5	>50	High

used (Smirnov, 1972). A comparison of these two newer scales has shown that they are, in general, no better than Maltsev's scale for estimating weediness (Tulikov, 1974).

An alternative approach is a visual estimation of the percentage cover of the total area by all or each species of weeds (Liberstein, 1973). These visual estimates of the degree of weediness are based on the five-point scale commonly used in phytosociological survey (Liberstein, 1973; Mueller-Dumbois and Ellenberg, 1974):

0   no weeds;
1   <10 per cent weed cover;
2   11–25 per cent weed cover;
3   26–50 per cent weed cover;
4   >50 per cent weed cover.

This method was developed for Soviet conditions by Khabibrakhmanov (1974) and now comprises a visual–qualitative estimate of the one-year weeds and a visual–quantitative estimate of the multi-year weeds and is applied when the major weeds are in blossom.

More recently in the USSR increased use has been made of biogeographical survey methods. These are based on counts and weights for both crops and weeds in small (typically $\leqslant 1 \text{ m}^2$) quadrats. An advantage of this approach is that the biological composition of the weeds, their distribution over the field and the phases of development as well as the degree of weediness can be estimated (Küchler and Zonneveld, 1988).

However, one set of features that is common to all of these techniques is that they are imperfect, subjective (to a greater or lesser extent) laborious and subject to sampling error. Consequently, the estimation of crop weediness from remotely sensed data has become an important issue in the USSR.

9.4. Remote sensing of weedy crops

To control weeds, up-to-date maps of the degree and location of weed-infested fields are required (Section 9.3). Maps of crop weediness in the USSR are less than ideal and only exist at large scales for selected fields. To improve this situation a technique for estimating and mapping crop weediness from remotely sensed data is required (Curran, 1985b; Ryerson and Curran, 1990). This has been developed for the USSR (Kondratyev and Fedchenko, 1982a; Kondratyev et al., 1986a) and is discussed in the remainder of this chapter.

9.4.1. Remote sensing of crop weediness in the earing phase

Spectral reflectance measurements of crops and weeds were used to develop a remote sensing approach to the assessment of crop weediness in the crops' earing phase. These measurements were made in the summer of 1978, when the crops were in the earing–blossoming phase and the weeds (e.g. ox-eye daisy, wild winter cress, shepherd's bag, wild radish, etc.) were in the blossoming phase. At earlier crop phases (tubing and bushing) the weeds and crops looked very similar.

Measurements of crop reflectance (SBC) were made so as to encompass much of the crops' variability; measurements were made for three kinds of rye (Kharhovskaya, Voskhod and Belta), four kinds of wheat (Mironovskaya-808, Illichevka, Kavkaz and Zarya), two kinds of barley (Moskovskiy-121 and Krasnoufimskiy) and one kind of oats (Hercules). In addition, the Mironovskaya-808 wheat had been subjected to different degrees of fertilizer application. The data for each crop and level of fertilizer application were averaged, and this led to a broadening of the confidence limits of the spectral-reflectance curves. This was necessary because it is very difficult to distinguish, simultaneously, between different varieties of a crop and to determine the degree of fertilizer application from remotely sensed data.

Homogeneous weed and crop canopies were identified and a set of SBC measurements for each was made from a height of 4 m. At this height the photometer's ground resolution element was approximately 1·4 m. Each measurement was acquired under strict observation conditions with the same observers using the same equipment under similar levels and geometries of illumination.

Measurements were made for eight types of weeds, with 5–15 measurements per type (Kondratyev and Fedchenko, 1979). These weeds were 85–90 per cent of those observed in crops growing in the non-chernozem zone. SBC measurements were also made for the wheat, rye, barley and oat crops mentioned previously. The 70 per cent confidence limits of the SBCs for the weeds and some crops overlapped (Figure 9.1). Although this would present a problem in classifying weed type, it is not important in mapping weediness because only the total weed coverage is required.

The crop SBCs were ratioed using

$$x_1 = \frac{r(\lambda = 0\cdot 84)}{r(\lambda = 0\cdot 65)} \qquad x_2 = \frac{r(\lambda = 0\cdot 55)}{r(\lambda = 0\cdot 65)}$$

$$x_3 = \frac{r(\lambda = 0\cdot 84)}{r(\lambda = 0\cdot 44)} \qquad x_4 = \frac{r(\lambda = 0\cdot 84)}{r(\lambda = 0\cdot 55)}$$

(9.1)

where $r(\lambda = 0\cdot 44)$ is blue reflectance, $r(\lambda = 0\cdot 55)$ is green reflectance, $r(\lambda = 0\cdot 65)$ is red reflectance and $r(\lambda = 0\cdot 84)$ is near-infrared reflectance.

The calculated ratios (Figure 9.2) were used to investigate the identification of weeds in winter rye crops. Winter rye and weeds display considerably different spectral reflectance properties (Figures 9.1 and 9.2). Consequently, it should be possible to develop a remote sensing technique to assess the degree of weediness in winter rye crops.

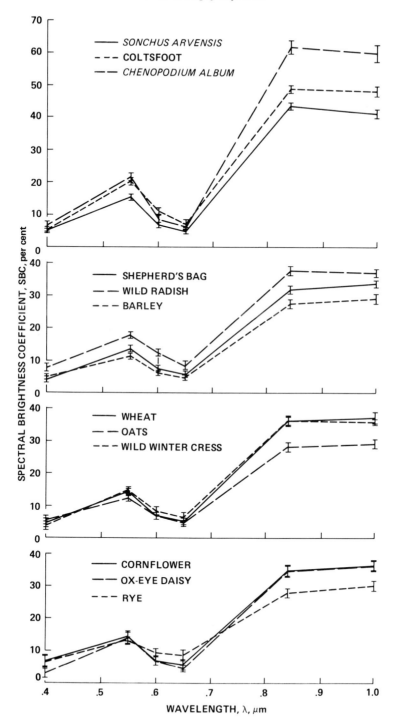

Figure 9.1 Spectral characteristics of a variety of weeds and crops.

Spectral reflectance of vegetation

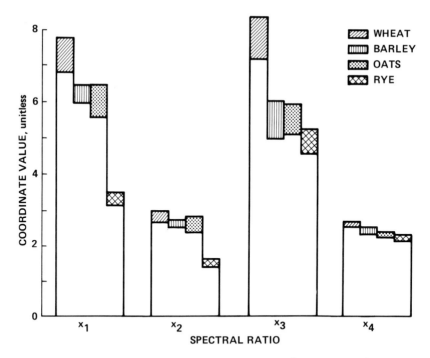

Figure 9.2 Spectral ratios for four crops. x_1 = *near-infrared/red;* x_2 = *green/red;* x_3 = *near-infrared/blue;* and x_4 = *near-infrared/green, and their 70 per cent confidence limits.*

SBC measurements of weed-infested fields were made with an airborne, multispectral, remote sensing system developed at the All-Union Research Institute for Agricultural Mechanical Engineering (Fedchenko and Kondratyev, 1981). This allowed the SBC to be measured in seven spectral regions over the $0 \cdot 4 - 1 \cdot 0$ μm wavelength interval. The reflected radiation was split into monochromatic signals by interference filters placed in a rotating holder, and was measured using an FEU-62 photomultiplier. The instrument had a 22° field-of-view and used MS-13 milk glass as a reference material. The spectral data recorded by this system, together with information on the time of measurement, illumination conditions and a weed survey, comprised the data set for the investigation.

Airborne SBC measurements of winter rye fields were made from an IL-14 aircraft. The altitude was dependent on the field size, and all observations were made under clear skies and with solar elevations of about 50°. Fifty-six fields, spanning a wide range of weediness, were analysed. These data were recorded when the rye was in the earing phase, because this is the period when the weed–crop differences are maximised.

Recognition of objects from measurements of SBC in several spectral regions is usually achieved by a statistical classification such as the Bayesian maximum likelihood classification. The following approach was adopted because the problem of recognition is an inverse problem. Suppose the given set of objects is confined to n classes. For some sets of spectral intervals m, a set of SBC values is prescribed for the objects of each ith class. Let elements Q_{ij} form the matrix $\hat{\mathbf{A}} = \{Q_{ij}\}_{i=1,j=1}^{mn}$ of $m \times n$ dimensionality.

Objects in the rural landscape (in this case fields) are seldom homogeneous. Consequently, the analysis should deal with objects representing a 'mixture' of ideal representatives of an initial set of objects from n classes with SBC $Q_j(\lambda_j)$ (Adams et al., 1986).

The unambiguous classification of a heterogeneous site is rarely possible as classes intergrade (Foody and Wood, 1987; Wood and Foody, 1989). Because reflected radiance is additive (the object's brightness is a sum of the brightnesses of its component parts), the following expression can be written for SBC $Q(\lambda)$ (measured in a fixed direction) of a site:

$$y(\lambda) = \sum_{i=1}^{n} x_i Q_i(\lambda) + \tilde{\varepsilon}(\lambda) \tag{9.2}$$

Here x_i is the portion of the percentage soil cover for the elements of the ith class and $\tilde{\varepsilon}(\lambda)$ is a random measurement error. With representatives of the chosen 'alphabet' of classes (assumed to embrace the whole variety of elements of real sites to be identified), the following conditions can be observed:

$$\sum_{i=1}^{n} x_i = 1, \qquad 0 \leqslant x_i \leqslant 1 \tag{9.3}$$

Suppose, for $\tilde{\varepsilon}(\lambda)$ the mathematical expectation $\bar{E}[\tilde{\varepsilon}(\lambda)] = 0$, and the respective covariance matrix had a diagonal of the form

$$\mathbf{K}_{\tilde{\varepsilon}} = \text{diag}\{\tilde{\sigma}_1^2, \ldots, \tilde{\sigma}_m^2\}$$

Equation 9.2 can be rewritten as a matrix

$$\mathbf{y} = \hat{\mathbf{A}}\mathbf{x} + \hat{\varepsilon} \tag{9.4}$$

and the condition (Equation 9.3) can be determined by a multitude of elements

$$S = \{x \in \mathbf{R}^n, \sum x_i = 1, x_i \geqslant 0\} \tag{9.5}$$

A linear multitude S can be determined with a set of the following scalar products for vectors $\mathbf{t}, \mathbf{C}_i, \ldots, \mathbf{C}_n$

$$\mathbf{t}_x^T = 1 \qquad \mathbf{C}_i^T x \geqslant 0 (i = 1, \ldots, n) \tag{9.6}$$

where T denotes transposition. Here $t^T = (1, \ldots, 1)$ and $\mathbf{C}_i$ is the ith line of the unit matrix.

Introducing into the consideration the function of errors

$$F(x) = \| \mathbf{y} - \hat{\mathbf{A}}\mathbf{x} \|^2 \mathbf{K}_{\tilde{\varepsilon}}^{-1} = (\mathbf{y} - \hat{\mathbf{A}}\mathbf{x})\mathbf{K}_{\tilde{\varepsilon}}^{-1}(\mathbf{y} - \hat{\mathbf{A}}\mathbf{x}) \tag{9.7}$$

gives a standard problem of minimising a square function (Equation 9.7) with linear restrictions (Equation 9.6). The presence of restrictions imposed by the inequalities in Equation 9.6 mitigates against a solution to the problem and requires the use of iterative algorithms.

In the approach reported here, however, the regular function

$$\tilde{\mathbf{G}}_{(x)} = \| \mathbf{y} - \hat{\mathbf{A}}\mathbf{x} \|^2 \mathbf{K}_{\tilde{\varepsilon}}^{-1} + S^2 \| \mathbf{x} - x_0 \|^2 \tag{9.8}$$

was used instead of Equation 9.7. Here x_0 is an initial approximation for the solution. Let the kth step of the process ($k = 0, 1, \ldots$) be completed. At the $(k + 1)$th step of the iterative process the problem is solved

$$\min_{\rho(k)} \tilde{\mathbf{G}}(x^{(k)} + \rho^{(k)})$$

with restrictions on the type of inequalities

$$\mathbf{C}_i^T \rho_i^{(k)} = 0 \quad (i = 1, \ldots, e) \qquad \mathbf{t}_{\rho_i^{(k)}}^T = 0 \tag{9.9}$$

Here e is the number of first restrictions—inequalities from Equations 9.6 that have been transformed into equalities in the $x^{(k)}$ point during a processing step in the iterative process.

Equation 9.9 has an apparent solution

$$\rho^{(k)} = \hat{\mathbf{H}}\mathbf{z} + \hat{\mathbf{H}}(x_0 - x^{(k)})$$

where

$$\mathbf{z} = (\hat{\mathbf{H}}\hat{\mathbf{A}}^T \mathbf{K}_{\tilde{\varepsilon}}^{-1} \hat{\mathbf{A}}\hat{\mathbf{H}} + S^2 \hat{\mathbf{I}})^{-1} \hat{\mathbf{H}}\hat{\mathbf{A}}_y^T$$

$\hat{\mathbf{H}}$ is the projector to a variety determined from Equation 9.9 and $\hat{\mathbf{I}}$ is the unit matrix.

The vector $\rho^{(k)}$ is the direction of descent at the kth step of the iterative procedure. Then, with the help of a one-dimensional search, the quantity of an optimal step may be found according to the condition

$$\alpha^{(k)} = \max \{\alpha : x^{(k)} + \alpha \rho^{(k)} \in S\}$$

In this case $x^{(k+1)} = x^{(k)} + \alpha^{(k)} \rho^{(k)}$ minimises $\mathbf{G}(x)$ in the S region. This procedure allows the solution of the problem in a finite number of steps.

This algorithm has been used to assess the degree of weediness of winter rye crops from spectral measurement data via the linear model

$$y = \hat{A}x + \tilde{\varepsilon}$$

where **y** is the vector of spectral measurement data, **x** is the vector to be estimated and $\hat{A}$ is a matrix composed of SBC of the jth type of plants corresponding to the ith measurement. The measurement conditions were such that the lines of the matrix were linearly dependent.

The portion of the percentage cover for winter rye is given in Table 9.2. In this table the first column is for the division of the estimated parameters into two classes—rye and weeds, where the SBCs for weeds were averaged over the different weed varieties in each spectral interval (Kondratyev and Fedchenko, 1980d). In the second and third columns the solutions were obtained for two and eight classes of weeds, respectively. The results show that with few spectral

Table 9.2 The percentage weediness of 56 winter rye fields derived from remotely sensed measurements.

Field number	Number of weed classes			Field number	Number of weed classes		
	1	2	8		1	2	8
1	6	38	27	29	8	7	10
2	0	0	0	30	4	6	9
3	14	43	35	31	59	73	78
4	21	48	30	32	58	72	66
5	14	44	62	33	73	67	65
6	100	100	100	34	62	53	47
7	29	53	74	35	6	31	18
8	23	50	1	36	30	0	12
9	46	64	68	37	36	24	24
10	0	0	0	38	39	31	26
11	90	93	93	39	73	77	68
12	0	0	0	40	15	26	16
13	2	0	0	41	0	30	36
14	64	25	0	42	100	100	100
15	100	100	100	43	0	0	0
16	4	37	25	44	100	100	100
17	31	0	0	45	0	0	0
18	57	27	33	46	0	0	0
19	1	35	27	47	100	100	100
20	45	64	70	48	19	47	34
21	37	59	68	49	30	54	16
22	10	9	12	50	100	100	100
23	100	100	100	51	85	75	74
24	32	0	0	52	0	0	0
25	0	0	0	53	9	40	36
26	100	100	100	54	6	38	38
27	100	100	100	55	18	46	50
28	0	0	0	56	28	53	57

measurements an increase of the dimensionality of the estimated **x** vector increases the complexity of the problem, which is manifested through an instability of the calculation scheme. This instability is apparent in the results given in Table 9.2. For approximately one third of the cases in Table 9.2 the calculated portion of the percentage cover for winter rye in the second and third columns is higher than in the first column. To remove this instability more measurements must be acquired. Results of measurements may also be linearly independent, and this may happen if, in the case of solving an inverse problem, there is some additional *a priori* information on possible *x* values.

This complication of the measurement procedures may not always be justifiable. The use of only two estimated quantities (rye and weeds) saves the calculations from instability, albeit at the expense of being less useful. This indicated the potential for numerical simulation experiments based on a two-class model for the determination of weediness in winter rye fields.

Forty $1 \cdot 5$ m $\times$ $1 \cdot 5$ m test sites in a field of winter rye were analysed to assess the accuracy of weediness estimation by remote sensing. At each site weediness was first estimated by a ground observer. Weediness was taken to be zero if the amount of winter rye was such that the soil and weeds could not be seen. Otherwise weediness was expressed as a percentage of the test site area. SBC measurements of the winter rye–weed system canopy were then made at each site. The measurements were obtained in the earing/milk-ripening phase of the winter rye, which is a period $1-1 \cdot 5$ months before harvesting, when weeds may grow to almost screen the soil.

The estimated weediness of these sites derived from the SBC measurements above closely corresponded to the degree of weediness determined by traditional means in the field (Figure 9.3). These studies have indicated that a possibility

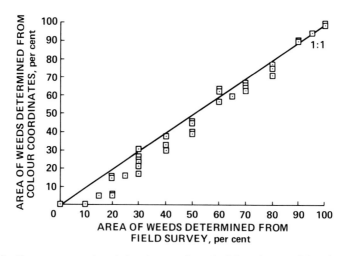

Figure 9.3 *Percentage cover of weeds in winter rye determined from the sum of the colour coordinates and by field survey.*

exists for the assessment of weediness in winter rye crops from the measurement of spectral reflectance made 1–1·5 months before harvesting. The technique requires refinement, and its applicability to this and other crops is now undergoing investigation (Kondratyev *et al.* 1986a).

9.4.2. Remote sensing of crop weediness during the wax-ripeness stage

Although the studies discussed above showed that the weediness of winter rye fields can be reliably estimated in the earing phase (about 1–1·5 months before harvesting), less-encouraging results have been found for other crops. With wheat, for instance, it is usually impossible to assess the degree of weediness in the earing phase, since the SBC of crops and weeds are similar and their confidence limits overlap (Figure 9.1). Attention therefore turned to the wax-ripeness state (2–3 weeks before harvesting) as at this period the crops tend to yellow, the weeds remain green and the SBC of the crops and weeds differ considerably (Figure 9.4). Typically, the crops' SBC did not have a maximum of green and a minimum of blue and red reflection, and the SBC of crops was much lower than that of the weeds in the 0·75–1·00 μm spectral interval.

The difference between the crop and weed SBC at the wax-ripeness stage, especially in red wavelengths, indicates an opportunity for the remote sensing of wheat weediness during this period. To substantiate this, the weed–crop system contrasts were calculated every 0·05 μm and the maximum contrast was found in the red waveband at 0·65 μm. The problem is essentially the solution of equation

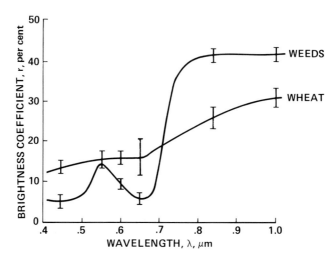

Figure 9.4 Spectral characteristics of weeds and wheat at the wax-ripeness stage. The mean square deviation of each measurement set is shown.

9.10 to find S_1

$$r_{cw} = S_1 r_c + S_2 r_w \qquad (9.10)$$

where r_{cw} is the SBC for the crop–weed mixture; r_c and r_w are the SBCs for crops and weeds, respectively; and S_1 and S_2 are the areas under crops and weeds, respectively.

Since

$$S_1 + S_2 = 1 \qquad (9.11)$$

Equation 9.10 can be rewritten as

$$S_1 = \frac{r_{cw} - r_w}{r_c - r_w} \qquad (9.12)$$

Substituting S_1 into Equation 9.11, S_2 can be obtained.

To calculate S_1 (Equation 9.12) it is necessary to know r_w, r_c and r_{cw}. The latter is derived from the remotely sensed data whereas the former two variables need to be known beforehand, usually from surface measurements. The accuracy of the calculation of S_1 clearly depends on how accurately r_w r_c and r_{cw} are determined. The accuracy of the latter depends on the precision of the instrument. The accuracy of the *a priori* data on r_c and r_w depend on their variability and the number of samples upon which their estimate is based.

Figure 9.4 shows that the SBC for weeds have a low mean-square deviation (MSD). Consequently, it is possible to use *in situ* measurements of r_w with confidence as its variability is low. However, the MSD of r_c at $0 \cdot 65$ μm is almost five times higher than for r_w. Consequently, the use of a mean r_c value in Equation 9.12 would lead to an inaccurate estimation of S_1.

The relatively high variability of r_c is thought to be a function of the gradual decline in the crop chlorophyll concentration. Tiny concentrations of chlorophyll ($0 \cdot 2$–$0 \cdot 3$ mg $(100 \text{ cm}^2)^{-1}$) can considerably alter the slope of the spectral reflectance curve in red wavelengths. Consequently, the interplant variations in chlorophyll concentrations as the crop goes through this wax-ripeness stage gives rise to a variable r_c. To improve the accuracy of the r_c estimate, measurements may have to be acquired for each specific region and date.

Crops and weeds are generally different heights. Often crops may be three to five times taller than weeds; although there are exceptions (e.g. *Sonchus arvenis* among barley and oats). Therefore, when the sensor views at nadir, crops, weeds and maybe even some soil may appear in its field-of-view. Increasing the off-nadir view angle generally results in the sensor 'seeing' more of the crop at the expense of the soil and weeds. Thus the r_c value can be reliably estimated from off-nadir data if the pointing precision of the instrument is adequate (Fedchenko, 1982c).

To determine if such an approach was possible measurements were taken with a lens photometer (Fedchenko and Kondratyev, 1981) which could view either at nadir or off-nadir (θ_r). This angle, θ_r, was selected to ensure that the weeds made no contribution to the observed reflectance of the sample sites. The angle was chosen so that $\theta_r < l/h$, where l is the distance between crops and h their height. As a guide for crops 70–100 cm tall with 10–20 cm tall weeds under standard sowing conditions (500 seeds m^{-2}) the angle θ_r should not be less than 40–50°.

Measurements made with the lens photometer from an aircraft showed that the SBC measured off-nadir was substantially dependent on the azimuth angle. Although this problem can be avoided with *in situ* measurements, by measuring at a fixed azimuth, this is not always possible for airborne measurements. To allow for this problem, a device which allowed simultaneous measurements both at nadir and at off-nadir angles was used.

With r_{cw} and r_c measured and r_w known, S_1 can be calculated from Equation 9.12. Since the acquisition of both the nadir and off-nadir measurements take only a second it is not necessary to estimate r_c for each point by measuring the target and reference and then calculating S_1. This cumbersome procedure can be

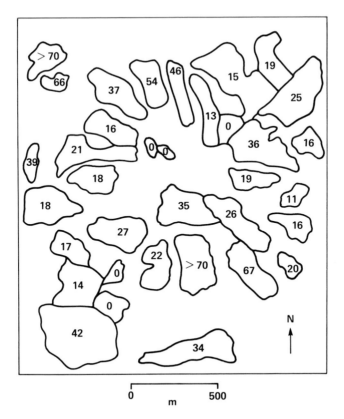

Figure 9.5 A map of crop weediness derived from remotely sensed data. Figures on the map indicate the percentage weediness.

substituted with the I_{cw}/I_c ratio, and S_1 derived from a previously prepared graph of the relationship I_{cw}/I_c on S_1.

If the angle θ_r is based on the current standard of sowing (500 seeds m^{-2}), then the weediness can only be estimated up to a cover of 50 per cent. At high weediness the weeds influence the brightness measured at θ_r, and reduce the accuracy of the estimated S_1. Approximately 70 per cent is the maximum weediness that can be determined by this technique and when weediness is higher observers should note that this is the case. Finally, the accuracy of the S_1 calculation will depend on other factors, such as wind or rain flattening (lodging) the crop.

Crop weediness has been estimated using this technique and remotely sensed data. For instance, the weediness of agricultural fields in the Kaluga region was estimated during the summer of 1980 using data recorded by a sensor on board a KA-26 helicopter at an altitude of 50–70 m. A total of 254 fields were sampled and all measurements were made when cloud cover was less than 30 per cent.

The results of such studies are most usefully presented as maps of weediness. However, this requires accurate geographical referencing. A map of weediness for a small territory of the Strekalovsky state farm of the Yukhnovsky district of the Kaluga region is shown in Figure 9.5. This map, based on reflectance spectra, characterises the weed coverage of the area and provides information to help to guide prompt agrotechnical and chemical weed-control measures. These maps can be produced very quickly, the example given in Figure 9.5 took only 1 h, from the time of measurement to delivery to users, and a map of the whole Kaluga region could be prepared in less than 2 days.

9.5. Concluding comments on the remote sensing of crop weeds

Remotely sensed data can be used to obtain prompt and accurate information on the weediness of some crops. By accelerating and rationalising the mapping process, fieldwork can be minimised and maps should be more timely and cheaper. The procedures are being developed and tested for other crops, and spectrophotometric instruments which permit in-flight data processing will help to speed up the procedure.

PART IV
Remote sensing of soils and crops from aircraft and satellites

10
Atmospheric correction of remotely sensed data

A major portion of the signal received by an airborne or spaceborne sensor is not from the Earth's surface but from the atmosphere. The magnitude of this atmospheric signal varies with wavelength, space and time, and is very difficult to measure or even approximate, yet such information is vital if the absolute signal from the Earth's surface is to be retrieved from the total signal received by the sensor. In the USSR a considerable research effort has been put into the atmospheric correction of remotely sensed data, using techniques drawn from across the physical sciences. This chapter discusses some of these techniques for use with airborne and satellite sensor data.

10.1. The influence of the atmosphere on remotely sensed data

Soil reflectance measurements were acquired simultaneously in November 1978 from an aircraft called the MGO (Main Geophysical Observatory) flying laboratory and from a Meteor-type experimental satellite (Curran, 1985a,c). The sensor onboard the MGO measured the spectral albedo and reflectance indicatrix at different heights and various solar elevations. Spectral albedo data were also derived from the digitized and atmospherically corrected, low spatial resolution, multispectral data recorded by the Moscow State University-type Meteor (MSU-M) radiometers onboard the Meteor-type satellite. The atmospheric correction included the calculation of radiation transfer in the surface–atmosphere system for the spectral channels of the MSU-M radiometers, to obtain the angular distributions of radiation as functions of angular coordinates and reflectances. The atmospheric transfer function represented a set of coefficients that approximated the atmospheric effect by taking into account the processes of interaction between radiation, gas and atmospheric aerosols.

In applying the atmospheric correction, account was taken of signal distortions by the satellite sensor and distortions caused by the transformation of radiation in the atmosphere. The radiometric correction was based on the use of a photometric wedge that was scanned and radio-transmitted during the backwards motion of the scanning radiometer. The albedo of individual pixels was retrieved in each spectral channel at given view angles and solar zenith angles using approximate coefficients of the atmospheric effect (obtained from solving radiative-transfer boundary-value problems).

The airborne system allowed the solution of inverse problems for retrieving reflectances. The sensors provided measurements at a variety of altitudes of relative angular brightness distributions (phase functions) in 10 spectral intervals (0·51–1·88 μm), and this allowed the calculation of spectral albedo. The spectral albedo was averaged by altitude to increase the measurement accuracy.

Relative surface phase functions at view zenith angles from nadir ($\theta_r > 0°$) were measured in each spectral interval in 12 azimuth directions relative to the Sun's vertical plane for flight paths over the Kara-Kum desert and agricultural fields of the Kherson region (Figure 10.1). An example of the observed angular-reflection anisotropy is illustrated for the desert by the relative (with

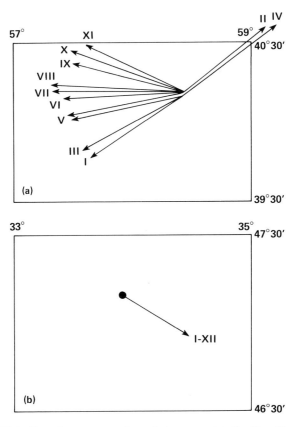

Figure 10.1 Flight lines for remote sensing missions over (a) the Kara-Kum desert and (b) agricultural fields for the Kherson region. Roman numerals represent different flying heights from I (low altitude, 200 m) to XII (high altitude, 8000 m).

Atmospheric correction of remotely sensed data

Table 10.1 Vertical distributions of volume attenuation coefficients (km^{-1}) for two channels of the Meteor sensor radiometers and two atmospheric models. Channel 1 is in visible wavelengths, channel 4 is in near-infrared wavelengths; model 1 is from the Leningrad State University model with aerosols and model 2 is for a Rayleigh atmosphere model without aerosols.

Atmospheric depth (km)	Channel 1 Model 1	Channel 1 Model 2	Channel 4 Model 1	Channel 4 Model 2
0–1	1.7746×10^{-1}	1.1453×10^{-2}	1.5851×10^{-1}	3.6936×10^{-2}
1–2	8.1209×10^{-2}	1.0412	8.9075×10^{-2}	2.8045
2–3	3.4065	9.2718×10^{-3}	4.1091	1.9599
3–4	2.4892	8.1606	2.7968	1.3203
4–5	2.1169	7.1268	2.1284	8.6010×10^{-3}
5–6	1.5087	6.1876	1.4007	5.6537
6–7	1.5559	5.3450	9.5259×10^{-1}	3.7642
7–8	8.4578×10^{-3}	4.5931	5.9413	2.5293
8–9	7.3339	3.9382	4.6807	1.7327
9–10	6.8600	3.3615	4.1881	1.2094
10–11	4.0837	2.8711	1.6769	8.5122×10^{-4}
11–12	2.9839	2.4467	9.2805×10^{-4}	6.1504
12–13	2.4691	2.0899	7.0989	4.6126
13–14	2.0896	1.8175	5.9709	3.6137
14–15	1.9190	1.5852	6.0935	2.9165
15–16	8.9511	1.3907	8.0135×10^{-3}	2.4055
16–17	9.4596	1.2409	8.6644	2.0136
17–18	1.0704×10^{-2}	1.1284	1.0035×10^{-2}	1.7001
18–19	1.3066	1.0451	1.2534	1.4470
19–20	1.2190	9.7070×10^{-4}	1.1687	1.2372
20–21	3.4833×10^{-4}	9.0195	2.7569×10^{-3}	1.0625
21–22	1.5847	8.3593	8.5860×10^{-4}	9.1427×10^{-5}
22–23	2.0829	7.6177	1.4291×10^{-3}	7.9094
23–24	1.3206	6.8561	6.9940×10^{-4}	6.8435
24–25	1.1656	6.1743	5.9642	5.9375
25–30	5.6456×10^{-5}	4.6877	1.3456	4.2258
30–35	3.6935	2.8838	1.0507	2.7534
35–40	2.2300	1.5502	8.1919×10^{-5}	1.7192
40–45	1.4273	8.0787×10^{-5}	6.8947	1.0484
45–50	8.8278×10^{-5}	4.0772	5.0922	6.2792×10^{-6}
50–70	4.0496	1.8503	2.3865	3.1970
70–100	9.9342×10^{-6}	8.1747×10^{-6}	3.2522×10^{-6}	1.5988
0–100	4.7370×10^{-1}	9.6023×10^{-2}	4.4085×10^{-1}	1.2577×10^{-1}

respect to nadir) brightness indicatrices in two spectral intervals for data collected from an altitude of 200 m (Figure 10.2).

In determining the atmospheric transfer function two models were used: first, a refractive index model for the atmospheric aerosol particles; and secondly, an aerosol size-distribution model (Kondratyev *et al.*, 1969). Mie theory (Slater, 1980) was used to calculate the aerosol's scattering and absorbing properties (specifically the cross sections of interaction and phase function) for these models. To estimate the aerosol effect, correction was also made for an idealized Rayleigh atmosphere without aerosols. The vertical distributions of the total-interaction cross sections between radiation and major optically active components for these atmospheric models are given in Table 10.1. Optical thickness was calculated by multiplying the data given in Table 10.1 by the estimated thickness of the respective atmospheric layers.

A comparison of the desert surface albedo derived from, first, airborne

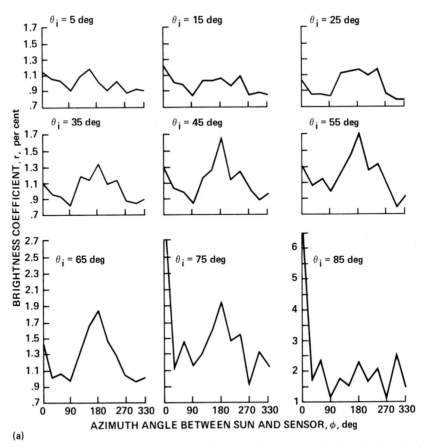

Figure 10.2 Brightness coefficient as a function of relative solar azimuth and view azimuth angle for the Kara-Kum desert: (a) $0·509$ μm and (b) $0·99$ μm.

remotely sensed measurements, and secondly, after atmospheric correction of the multispectral data from the satellite sensor, revealed some considerable differences (Figure 10.3).

The desert albedo measurements had been made from an altitude of 200 m at the wavelengths $0 \cdot 509$, $0 \cdot 701$, $0 \cdot 960$ and $0 \cdot 990$ μm. The agricultural albedo measurements were made at a wavelength of $0 \cdot 990$ μm from an altitude of 300 m. Satellite sensor data comprised four MSU-M channels for the desert and the near-infrared channel 4 for the agricultural region.

The results showed that changes in spectral reflectance for the desert did not exceed 15 per cent after atmospheric correction. Changes in excess of 50 per cent were, however, observed in the satellite sensor's channel 4 data for agricultural fields. These could be accounted for by a low accuracy of the radiometric correction and the averaging of radiance in channel 4 over a broad waveband. Furthermore, the areal averaging of measurements that takes place in low-spatial-resolution satellite sensor data ignores the heterogeneity of the surface and its effect on measured radiance.

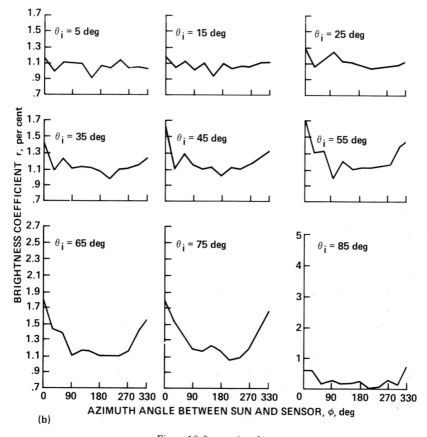

Figure 10.2 continued

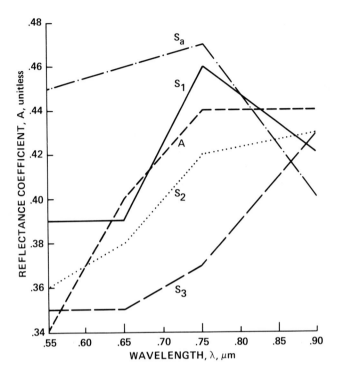

Figure 10.3 Surface albedo from both aircraft sensor measurements and atmospherically corrected satellite sensor data. A, desert albedo, $H_f = 200$ m; S_a, surface–atmosphere system albedo from Meteor sensor data; S_1 and S_2 surface albedo after atmospheric correction with use of models 1 and 2, respectively; and S_3 the same as for S_1 and S_2, but without account of reflectance anisotropy.

Some changes in the observed reflectance (Figure 10.3) were due to the inadequacy of the atmospheric aerosol model and the lack of correction for reflectance anistropy. Therefore, to undertake reliable experiments utilising remotely sensed data, the optical properties of the atmosphere need further examination and more accurate quantitative description. The results show that ignoring atmospheric effects gives rise to discrepancies of more than 40 per cent between airborne and spaceborne reflectance measurements. Experience has shown that the use of a set of approximation coefficients for atmospheric effects of a Rayleigh model atmosphere can reduce this error to about 15 per cent and, by adopting the Leningrad State University aerosol model for atmospheric correction, errors of less than 10 per cent can be achieved. Thus, determining the atmospheric transfer function, in terms of optical atmospheric aerosol models, enhances the accuracy with which remotely sensed data, recorded from satellite sensors, can be analysed. However, to do this the optical properties of the atmosphere should be assessed at the time of data acquisition.

10.2. Techniques for the atmospheric correction of remotely sensed data

In the USSR two major areas of investigation are associated with the atmospheric correction of data from satellite sensors. First, theoretical consideration of the main laws of formation for the outgoing radiation field, and secondly, the practical application of atmospheric radiation transfer models in processing digitised remotely sensed data (Kondratyev et al., 1975).

The theory of the interrelationship between brightness in the atmosphere and on the Earth's surface (atmospheric transfer function; Kondratyev and Smoktiy, 1972, 1973) has served as a mathematical basis for assessing the atmospheric effect on the brightnesses and brightness contrasts of natural targets observed from spaceborne remote sensors.

The actual atmospheric correction applied during data processing is usually based on simplified models of radiation transfer. Emphasis has typically been placed on analysing uncorrected and corrected data and checking that the removal of the effect of atmospheric haze (i.e. an increase in brightness) intensifies the contrasts between terrestrial features.

The theoretical basis of radiation transfer has also been developed and refined, using information from other fields such as nuclear physics (Dave and Canosa, 1974; Dave, 1975a,b, 1980). However, simplified models that do not require the cumbersome solution of boundary-value problems that are inherent in radiation transfer theory are usually used in the processing of remotely sensed data in the USSR.

Models for the atmospheric correction of remotely sensed data (Rogers, 1973; Switzer et al., 1981) often take the measurement data to be a sum of two components; surface brightness (useful signal) and atmospheric brightness (atmospheric noise). This allows, with additional measurements (Peacock, 1974; Griggs, 1975), the assessment of atmospheric brightness and the retrieval of surface brightness by subtraction. The useful signal is generally represented multiplicatively and a free term describes the brightness variations not considered by the correction models. The multiplicative constant is determined by the topography of the site and the Sun's position in the sky, and a factor for the constant is in direct proportion to average reflectance (Asmus et al., 1980). Such atmospheric correction models are useful in giving a simplified parameterization of the major processes of the formation of the outgoing radiation field. However, in the USSR experience has shown that the basis for such a parameterization should follow from detailed knowledge of the interaction of radiation with the Earth–atmosphere system and not from intuition.

Approaches to atmospheric correction of digitized remotely sensed data in the USSR can now be considered (Kozoderov, 1977; Kozoderov and Shulgina, 1980). Atmospheric correction is based on the solution of the following boundary-value problems:

(1) for the mean diffuse radiance with a spectral interval

$$\mu \frac{\partial \bar{I}(z,\mu,\phi)}{\partial z} + \sigma_t(z)\bar{I}(z,\mu,\phi)$$

$$= \sum_{p'=1}^{p'} \sigma_s^{p'}(z) \sum_{p'=0}^{\infty} b_k^{p'}(z) \frac{2k+1}{4\pi} \left[\int_0^{2\pi} d\phi' \int_{-1}^{1} d\mu' \bar{I}(z,\mu',\phi') P_k(\mathbf{s},\mathbf{s}') \right.$$

$$+ S_0 P_k \left[\mu \cos \tilde{z} + \sqrt{1-\mu^2} \sqrt{1-\cos^2 \tilde{z}} \cos(\phi - \phi_\odot) \right]$$

$$\left. \times \exp\left(-\frac{1}{|\cos \tilde{z}|} \int_z^H \sigma_t(z') \, dz' \right) \right] \tag{10.1}$$

$$\bar{I}(H, \mu^-, \phi) = 0$$

$$\bar{I}(0, \mu^+, \phi) = D_0 \bar{R}(\mu^+, \phi, \cos \tilde{z}, \phi_\odot)$$

$$+ \int_0^{2\pi} d\phi' \int_{-1}^{0} \bar{I}(0, \mu^-, \phi') \bar{R}(\mu^+, \phi, \mu^-, \phi') |\mu^-| \, d\mu^-$$

(2) for the fluctuation component of a Fourier series of radiance within the same spectral interval

$$\mu \frac{\partial I'(z, \mathbf{p}, \mathbf{s})}{\partial z} + [\sigma_t(z) - i(\mathbf{p}, \mathbf{s})] I'(z, \mathbf{p}, \mathbf{s})$$

$$= \sum_{p'=1}^{p'} \sigma_s^{p'}(z) \int_\Omega I'(z, \mathbf{p}, \mathbf{s}) \gamma_{p'}(z, \mathbf{s}, \mathbf{s}') \, d\mathbf{s}'$$

$$I'(H, \mathbf{p}, \mathbf{s}) = 0 \qquad \mathbf{s} \in \Omega_- \tag{10.2}$$

$$I'(0, \mathbf{p}, \mathbf{s}) = \int_{\Omega_-} \bar{R}(\mathbf{s}, \mathbf{s}') I'(0, \mathbf{p}, \mathbf{s}') \, d\mathbf{s}' + \int_{\Omega_-} R'(\mathbf{s}, \mathbf{s}') \bar{I}(0, \mathbf{s}') \, d\mathbf{s} \qquad \mathbf{s} \in \Omega_+$$

Here $I(z, \mu, \phi) = \bar{I} + I'$, and the coefficient of surface radiance

$$R(\mu^+, \phi, \mu^-, \phi^-) = \bar{R} + R',$$

therefore the combined correlation moment $R'I'$ in Equation 10.2 can be neglected. The remaining notations in Equations 10.1 and 10.2 correspond to those assumed in the theory of neutron transfer, and are therefore not the same as those used in Figure 1.1: where (μ, ϕ) is the zenith view angle cosine and view angle azimuth; $(\cos \tilde{z}, \phi_\odot)$ is the zenith solar angle cosine and solar azimuth; z is the vertical coordinate measured upwards along a normal to the surface; H is the top of the atmosphere; (μ^+, μ^-) are, respectively, positive and negative values of

μ; (Ω_+, Ω_-) are, respectively, the upper and bottom hemisphere. The term $s = (\mu, \sqrt{1-\mu^2}\cos\phi, \sqrt{1-\mu^2}\sin\phi)$ is for the macroscopic scattering cross section ($\delta_s^{p'}$), phase function ($\gamma_{p'}$) and coefficients of the Legandre polynomial expansion ($b_k^{p'}$) in series P_k from the scattering angle ss' for each of the p' optically active atmospheric components, in which number P' (atmospheric aerosols and gases); $\delta_t = \Sigma_{p'=1}^{p'} \delta_s^{p'} + \delta_\varepsilon^{p'}$ is the total cross section ($\delta_\varepsilon^{p'}$ is the absorption cross section); $\mathbf{p} = (p_x, p_y)$ is a vector of spatial frequencies and S_0 is the solar constant for the spectral interval $\Delta\lambda$

$$D_0 = S_0 |\cos \tilde{z}| \exp\left(-\frac{1}{|\cos \tilde{z}|}\int_0^H \sigma_t(z')\,dz'\right)$$

A representation of the second of the boundary-value problems in Equation 10.2 could be used as a two-dimensional generalisation of the solution for one-dimensional boundary-value problems (Equation 10.1).

One possible approach to solving boundary-value problems (Equations 10.1 and 10.2) is the Legandre polynomial expansion in a series using the spherical harmonics method

$$\bar{I}(z, \mu, \phi) \approx$$

$$\sum_{n=0}^{2N+1} \frac{2n+1}{4\pi} \Psi_{n,0}(z) P_n(\mu) + 2 \sum_{n=1}^{2N+1} \sum_{m=1}^{\min(n,M)} \frac{2n+1}{4\pi} \frac{(n-m)!}{(n+m)!} P_n^m(\mu)$$

$$\times [X_{n,m}(z)\cos m\phi - Y_{n,m}(z)\sin m\phi] \quad (10.3)$$

where $X_{n,m}(z) = \mathrm{Re}\,\Psi_{n,m}(z)$; $Y_{n,m}(z) = \mathrm{Im}\,\Psi_{n,m}(z)$, and N and M are boundary indices of the number of spherical harmonics considered (an odd number of the series terms is explained by a representation of boundary conditions in Marshak form discussed below). Consequently, the problem reduces to solving a system of ordinary differential equations relative to vectors of the angular distribution moments for radiance ($n = 0, 1, \ldots, 2N+1$; $m = 0, 1, \ldots, \min(n, M)$)

$$(n+1-m)\frac{d\begin{pmatrix}X_{n+1,m(z)}\\Y_{n+1,m(z)}\end{pmatrix}}{dz} + (n+m)\frac{d\begin{pmatrix}X_{n-1,m(z)}\\Y_{n-1,m(z)}\end{pmatrix}}{dz}$$

$$+ \sigma_t(z)[1-c_n(z)](2n+1)\begin{pmatrix}X_{n,m},(z)\\Y_{n,m},(z)\end{pmatrix}$$

$$= S_0 \exp\left(-\frac{1}{|\cos\tilde{z}|}\int_z^H \sigma_t(z')\,dz'\right)\sigma_t(z)c_n(z)(2n+1)$$

$$\times P_n^m(|\cos\tilde{z}|)(-1)^{n-m}\begin{pmatrix}\cos m\phi_\odot\\-\sin m\phi_\odot\end{pmatrix} \quad (10.4)$$

where

$$\sigma_t(z)c_n(z) = \sum_{p'=1}^{p'} \sigma_s^{p'}(z) b_n^{p'}(z)$$

Upon the finite-difference presentation of the systems in Equation 10.4 the following systems of three-diagonal algebraic equations in a vector–matrix form can be obtained, for each of j atmospheric layers:

$$\widetilde{\mathbf{T}}_j^{(1)} \boldsymbol{\psi}(z_j) = \widetilde{\mathbf{T}}_j^{(2)} \boldsymbol{\psi}(z_{j-1}) + \mathbf{Q}^{(j)} \tag{10.5}$$

where $\boldsymbol{\psi}$ represents the matrix lines

$$\begin{pmatrix} \psi_{0,0} & \psi_{1,0} & \psi_{2,0} & \cdots & \psi_{2N-1,0} & \psi_{2N,0} & \psi_{2N+1,0} \\ 0 & \begin{Bmatrix} X \\ Y \end{Bmatrix}_{1,1} & \begin{Bmatrix} X \\ Y \end{Bmatrix}_{2,1} & \cdots & \begin{Bmatrix} X \\ Y \end{Bmatrix}_{2N-1,1} & \begin{Bmatrix} X \\ Y \end{Bmatrix}_{2N,1} & 0 \\ 0 & 0 & \begin{Bmatrix} X \\ Y \end{Bmatrix}_{2,2} & \cdots & \begin{Bmatrix} X \\ Y \end{Bmatrix}_{2N-1,2} & 0 & 0 \\ \cdots & \cdots & \cdots & \cdots & \cdots & \cdots & \cdots \\ 0 & 0 & \begin{Bmatrix} X \\ Y \end{Bmatrix}_{N,N} & \begin{Bmatrix} X \\ Y \end{Bmatrix}_{N-1,N} & 0 & 0 \end{pmatrix} \begin{matrix} m=0 \\ m=1 \\ m=2 \\ \vdots \\ m=N \end{matrix}$$

$\mathbf{T}_j^{(1)}$ and $\mathbf{T}_j^{(2)}$ are matrices of the left-hand parts of the finite-difference equations (and are analogues to Equation 10.4) and $\mathbf{Q}$ represents vectors of the right-hand parts.

From continuity conditions for vectors $\boldsymbol{\psi}$ on the inner boundaries of individual atmospheric layers, with the optical characteristics for every m constant within their limits, the following relationships between even and odd components at the top and bottom of the atmosphere are obtained

$$\begin{aligned} \boldsymbol{\psi}_{\text{even}}(H) &= \hat{p}_1 \boldsymbol{\psi}_{\text{even}}(0) + \hat{p}_2 \boldsymbol{\psi}_{\text{odd}}(0) + \mathbf{Q}_{\text{even}} \\ \boldsymbol{\psi}_{\text{odd}}(H) &= \hat{p}_3 \boldsymbol{\psi}_{\text{even}}(0) + \hat{p}_4 \boldsymbol{\psi}_{\text{odd}}(0) + \mathbf{Q}_{\text{odd}} \end{aligned} \tag{10.6}$$

where $\hat{p}_1$, $\hat{p}_2$, $\hat{p}_3$ and $\hat{p}_4$ matrix elements and vectors $\mathbf{Q}_{\text{even}}$, $\mathbf{Q}_{\text{odd}}$ are found from a specially selected system of orthogonal vectors with the use of a fast solution of three-diagonal algebraic equations. The latter enabled a considerable improvement of the efficiency of this refined version of the spherical-harmonics method relative to the classical method (Smelov, 1978), and it has been reported within the USSR to be accurate and to be computationally rapid.

The vector–matrix form of boundary-conditions for the one-dimensional problem (Equation 10.1) in Marshak form for Lambertian reflection ($\bar{R}, = \bar{A}/\pi$, where $\bar{A}$ is the albedo) is written as

$$\psi_{odd}(H) = \hat{p}_5 \psi_{even}(H) \psi_{odd}(0) = \left(-\hat{p}_5 + \frac{4\bar{A}}{1+\bar{A}} \hat{p}_6\right) \psi_{even}(0) + D_0 \frac{4\bar{A}}{1+\bar{A}} R_0 \qquad (10.7)$$

where matrices $\hat{p}_5$, $\hat{p}_6$ and the vector R_0 are determined from integrals or Legendre polynomials.

A solution for boundary-value problem (Equation 10.1) follows from Equations 10.6 and 10.7 for each of the azimuth harmonics

$$\psi_{even}(0) = \hat{p}_7^{-1}\left(-\frac{4\bar{A}}{1+\bar{A}} D_0(\hat{p}_1 - \hat{p}_5\hat{p}_2)R_0 - (Q_{odd} - \hat{p}_5 Q_{even})\right) \qquad (10.8)$$

where

$$\hat{p}_7 = (\hat{p}_3 - \hat{p}_5\hat{p}_1) + (\hat{p}_4 - \hat{p}_5\hat{p}_2)\left(-\hat{p}_5 + \frac{4\bar{A}}{1+\bar{A}}\hat{p}_6\right)$$

A generalisation of the lower boundary conditions for Equation 10.1 in the case of non-orthotropic reflection

$$\left(1 - \frac{4\pi\bar{R}}{1+d}\right)\begin{Bmatrix}X(0)\\Y(0)\end{Bmatrix}_{2n-1+m,m} = \hat{p}_5^{(m)}\begin{Bmatrix}X(0)\\Y(0)\end{Bmatrix}_{2n-2+m,m} \frac{4\pi\bar{R}}{1+d}\hat{p}_6^{(m)}\begin{Bmatrix}X(0)\\Y(0)\end{Bmatrix}_{2n-2+m,m}$$

$$+ 4\pi^2 \frac{\bar{R}D_0}{1+d} R_s^{(m)}\begin{Bmatrix}1\\-1\end{Bmatrix} \qquad (10.9)$$

follows from a presentation of a two-parameter family of reflection indicatrices (d, g parameters) in the form of Henye–Greenstein indicatrices

$$\frac{d}{1+d}\left(1 + \frac{1-g^2}{d[1-g^2-2g(ss')]^{3/2}}\right) = \frac{d}{1+d}\left[1 + \frac{1}{d}\sum_{n=0}^{\infty}\left(P_n(\mu^+)P_n(\mu^-)\right.\right.$$

$$\left.\left. + 2\sum_{l=1}^{n}\frac{(n-l)!}{(n+l)!}P_n^l(\mu^+)P_n^l(\mu^-)\cos l(\phi-\phi')\right)(2n+1)g^n\right]$$

In Equation 10.9 the R_s^m vector depends on the parameter g of Legendre polynomials and standard integrals of these polynomials and $\hat{p}_{5,6}^m$ are generalised matrices of $\hat{p}_{5,6}$. Calculation schemes for solving the more-complex boundary-layer problem (Equation 10.1) are described in Kozoderov et al. (1980a).

The solution of this boundary-value problem (Equation 10.1) can be generalised for the case of the Earth's heterogeneous surface. In particular, the equation for the boundary-value problem (Equation 10.2) can be written in the vector–matrix form

$$\psi(H) = \hat{p}\psi(0) \tag{10.10}$$

where the ψ vector (matrix line) becomes 'stretched' in its number of dimension $2(2N + 1)N$

$$\psi = \{\psi^{Re}, \tilde{\psi}^{Im}\} = \{\psi_{0,0}^{Re}; \psi_{1,0}^{Re}; \ldots; \psi_{2N+1,0}^{Re}; \ldots; \psi_{m,m}^{Re}; \ldots; \psi_{2N+1-m,m}^{Re}; \ldots;$$

$$\psi_{N,N}^{Re}; \psi_{N+1,N}^{Re}; \tilde{\psi}_{0,0}^{Im}; \tilde{\psi}_{1,0}^{Im}; \ldots; \tilde{\psi}_{2N+1,0}^{Im}; \ldots; \tilde{\psi}_{m,m}^{Im}; \ldots; \tilde{\psi}_{2N+1-m,m}^{Im}; \ldots; \tilde{\psi}_{N,N}^{Im}; \tilde{\psi}_{N+1,N}^{Im}\}$$

and the matrices $\hat{p}_1$, $\hat{p}_2$, $\hat{p}_3$ and $\hat{p}_4$ become cellular

$$\hat{p} = \begin{pmatrix} \hat{p}_1 & \hat{p}_2 \\ \hat{p}_3 & \hat{p}_4 \end{pmatrix}$$

The boundary conditions here become

$$\begin{array}{ll} \psi_{even}^{Re}(H) = -\hat{p}_{51}\psi_{odd}^{Re}(H) & \tilde{\psi}_{even}^{Im}(H) = -\hat{p}_{51}\tilde{\psi}_{odd}^{Im}(H) \\ \psi_{even}^{Re}(0) = -\hat{p}_{61}\psi_{odd}^{Re}(0) + \widetilde{\mathbf{R}}_0 & \tilde{\psi}_{even}^{Im}(0) = -\hat{p}_{61}\tilde{\psi}_{odd}^{Im}(0) + \widetilde{\widetilde{\mathbf{R}}}_0 \end{array} \tag{10.11}$$

where

$$\hat{p}_{51} = \begin{pmatrix} \hat{p}_5 & 0 \\ 0 & \hat{p}_5 \end{pmatrix} \qquad \hat{p}_{61} = \begin{pmatrix} \hat{p}_6 & 0 \\ 0 & \hat{p}_6 \end{pmatrix}$$

and the solution for Equation 10.2 comes to

$$\begin{array}{l} \psi_{even}^{Re}(0) = -\hat{p}_7^{-1}(\hat{p}_4 - \hat{p}_{51}\hat{p}_2)\widetilde{\mathbf{R}}_0 \\ \tilde{\psi}_{even}^{Im}(0) = -\hat{p}_7^{-1}(\hat{p}_4 - \hat{p}_{51}\hat{p}_2)\widetilde{\mathbf{R}}_0 \end{array} \tag{10.12}$$

This calculation scheme and its application to remote sensing is discussed in detail by Kozoderov and Mishin (1980).

Results of calculations, using Equations 10.1 and 10.2 and radiance and normalised to the solar constant, for the spectral interval $0 \cdot 5 - 0 \cdot 55$ μm are illustrated in Figures $10 \cdot 4 - 10 \cdot 6$.

Variations in the brightness, I, of a moderately turbid atmosphere as a function of the survey geometry (μ^+, cos $\check{z}$, $\phi - \phi_\odot$) and the albedo of an homogeneous Lambertian surface are shown in Figure 10.4. The dependence of $\bar{I}$ on μ^+ coincides with the dependence of I on cos $\check{z}$, based on the optical theorem of Born and Wolf (1973). Figure 10.4 also shows the dependence of I on the

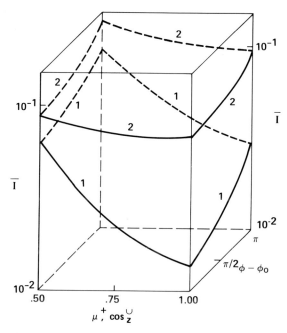

Figure 10.4 The relationships between mean brightness, sensor angles and mean surface brightness.

angular coordinates of scanning sensor systems, for which applied plane-parallel geometry holds. For each of k models of atmospheric stratification and each of j types of reflection non-orthotropicity $\bar{A}_j$, data similar to Figure 10.4, could be approximated by the following system of Chebyshev square polynomials:

$$\bar{I}_{kj}(\mu^+, \phi - \phi_\odot, |\cos \check{z}|, \bar{A}_j) \approx a_1 + a_2 \hat{T}_1(\bar{A}_j)$$
$$+ a_3 \hat{T}_1(|\cos \check{z}|) + a_4 \hat{T}_1(\mu^+) + a_5 \hat{T}_1(\phi - \phi_\odot) + a_6 \hat{T}_2(\bar{A}_j)$$
$$+ a_7 \hat{T}_1(\bar{A}_j)\hat{T}_1(|\cos \check{z}|) + a_8 \hat{T}_1(\bar{A}_j)\hat{T}_1(\mu^+) + a_9 \hat{T}_1(\bar{A}_j)\hat{T}_1(\phi - \phi_\odot)$$
$$+ A_{10} \hat{T}_2(|\cos \check{z}|) + a_{11} \hat{T}_1(|\cos \check{z}|)\hat{T}_1(\mu^+) + a_{12} \hat{T}_1(|\cos \check{z}|)\hat{T}_1(\phi - \phi_\odot)$$
$$+ a_{13} \hat{T}_2(\mu^+) + a_{14} \hat{T}_1(\mu^+)\hat{T}_1(\mu^+)(\phi - \phi_\odot)a_{15} \hat{T}_2(\phi - \phi_\odot) \qquad (10.13)$$

Here

$$T_1(x) = 2x - 1 \qquad T_2(x) = 2(2x - 1)^2 - 1$$

Table 10.2 gives coefficients a_n ($n = 1, ..., 15$) for the mean-statistical model characterised by Figure 10.4.

Figure 10.5 indicates the effect of atmospheric state and a homogeneous non-orthotropic surface on the nadir brightness radiance at $\cos \check{z} = 0 \cdot 7$. The data illustrated in Figure 10.5 are useful in the choice of models (indices k and j) for

Table 10.2 *Coefficients for approximating atmospheric effects by Chebyshev orthogonal polynomials using four intervals of mean albedo ($\bar{A}$) from 1 to 100 per cent.*

	$\bar{A}$ (per cent)			
a_i	1–10	10–30	30–50	50–100
a_1	$5 \cdot 679 \times 10^{-3}$	$6 \cdot 122 \times 10^{-3}$	$6 \cdot 569 \times 10^{-3}$	$7 \cdot 823 \times 10^{-3}$
a_2	$4 \cdot 669$	$4 \cdot 796$	$4 \cdot 947$	$4 \cdot 505$
a_3	$4 \cdot 907$	$5 \cdot 186$	$5 \cdot 052$	$5 \cdot 120$
a_4	$-9 \cdot 228 \times 10^{-4}$	$-1 \cdot 268$	$-1 \cdot 894$	$-2 \cdot 971$
a_5	$-4 \cdot 696 \times 10^{-5}$	$-9 \cdot 367 \times 10^{-5}$	$-1 \cdot 420 \times 10^{-4}$	$-2 \cdot 682 \times 10^{-4}$
a_6	0	$7 \cdot 208$	$-2 \cdot 220 \times 10^{-5}$	$2 \cdot 630$
a_7	$4 \cdot 805 \times 10^{-3}$	$4 \cdot 918 \times 10^{-3}$	$5 \cdot 063 \times 10^{-3}$	$5 \cdot 882 \times 10^{-3}$
a_8	$-2 \cdot 449 \times 10^{-4}$	$-7 \cdot 708 \times 10^{-5}$	$-3 \cdot 020 \times 10^{-4}$	$-2 \cdot 842 \times 10^{-4}$
a_9	$-1 \cdot 224 \times 10^{-5}$	$-1 \cdot 715$	$-1 \cdot 274 \times 10^{-5}$	$-1 \cdot 011 \times 10^{-6}$
a_{10}	0	0	0	0
a_{11}	$1 \cdot 257 \times 10^{-4}$	$-4 \cdot 348 \times 10^{-5}$	$1 \cdot 082 \times 10^{-4}$	$-2 \cdot 116 \times 10^{-4}$
a_{12}	$-3 \cdot 491 \times 10^{-6}$	$-9 \cdot 052 \times 10^{-6}$	$-5 \cdot 258 \times 10^{-6}$	$-1 \cdot 504 \times 10^{-5}$
a_{13}	$9 \cdot 732 \times 10^{-5}$	$2 \cdot 030 \times 10^{-4}$	$3 \cdot 121 \times 10^{-4}$	$5 \cdot 839 \times 10^{-4}$
a_{14}	$9 \cdot 173 \times 10^{-6}$	$2 \cdot 126 \times 10^{-6}$	$3 \cdot 373 \times 10^{-5}$	$6 \cdot 801 \times 10^{-5}$
a_{15}	$1 \cdot 334$	$3 \cdot 285$	$5 \cdot 235 \times 10^{-6}$	$1 \cdot 207$

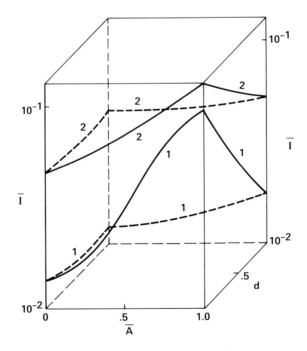

Figure 10.5 *The relationship between mean surface brightness for (1) transparent and (2) turbid atmospheres;* $d = 0$ *is a Lambertian and* $d = 1$ *is a specular surface.*

which the approximation in Equation 10.3 is needed. An approximation permits a real-time solution for each pixel of a satellite sensor image, of the inverse problem of retrieving $\bar{A}_j$ for measured mean radiance I and known angular survey coordinates (μ^+, cos $\tilde{z}$, $\phi - \phi_\odot$); known coefficients a_n; a given model of atmospheric turbidity (k index) and a known factor of reflection non-orthotropicity (j index).

Figure 10.6 illustrates the dependence of brightness on the surface distribution of natural objects, characterised by mean albedo $\bar{A}$ and spatial frequency p_x, with variable atmospheric state. The atmosphere is a low-frequency filter, smoothing the high-frequency component of the spectrum. The respective boundary frequency therefore decreases with increasing atmospheric turbidity τ (Figure 10.6).

Using the mathematical formalism of linear systems theory (Feldbaum and Butkovsky, 1971), in which the fluctuation component of radiance is represented as

$$I'(p_\xi, p_\eta) = |\mathbf{I}'(p_\xi, p_\eta)| e^{i \arg I'(p_\xi, p_\eta)}$$

the solution for the boundary-value problem (Equation 10.2) can be written as

$$I_k(\xi, \eta) = \frac{1}{(2\pi)^2} \iint_{-\infty}^{+\infty} A'(p_\xi, p_\eta) \hat{I}_k(p_\xi, p_\eta) e^{-i(\xi p_\xi + \eta p_\eta)} \, dp_\xi \, dp_\eta \quad (10.14)$$

where the spatial frequency characteristic of the atmospheric for brightness $\hat{I}_k$ is presented in the form of simple, but cumbersome, expressions of integrals in which the subintegral functions are determined by series of Legandre polynomials and by some similar functions of spatial frequencies, angular coordinates and atmospheric parameters. These latter functions are expressed in terms of solutions (Equation 10.7) to the boundary-value problem (Equation 10.1) for asymptotic values of infinitesimal and high frequencies, as well as more-general solutions (Equation 10.12) to the boundary-value problem (Equation 10.2). These expressions and their elaborations can be found in Kozoderov and Mishin (1980). A similar mathematical formalism, but with an iterative algorithm for determining $\hat{I}_k$, has been used in analysing the amplitude–frequency characteristic $|I'|$ (Mishin and Sushkeich, 1980).

Equation 10.14 can be used to retrieve the fluctuation component of the surface albedo

$$A'(x, y) = \frac{1}{(2\pi)^2} \iint_{-\infty}^{+\infty} I'(\xi, \eta) \, dp_\xi \, dp_\eta \, d\xi \, d\eta$$

$$\iint_{-\infty}^{+\infty} \frac{\exp\{i[(\xi - x)p_\xi + (\eta - y)p_\eta] - i \arctan(\operatorname{Im} I'/\operatorname{Re} I')\}}{|\hat{I}(p_\xi, p_\eta)|} \quad (10.15)$$

However, in constrast to the inversion of Equation 10.13, of the problem in

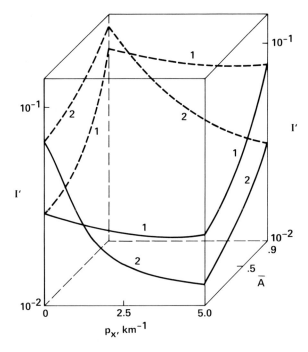

Figure 10.6 The relationship between the spatial frequency of mean surface brightness and atmospheric turbidity. (1) Transparent atmosphere and (2) turbid atmosphere.

Equation 10.1 (in which the inverse problem is reduced to the solution of a square equation relative to $\bar{A}_j$), problem (Equation 10.15) for A' is incorrect (Tikhonov and Arsenin, 1979). A regularised solution to the latter problem may be written as (Goncharsky et al., 1971)

$$A'_\alpha(x, y) = \iint_{-\infty}^{+\infty} \Pi_\alpha(x - \xi, y - \eta) I'(\xi, \eta) \, d\xi \, d\eta \qquad (10.16)$$

where

$$\Pi_\alpha(x - \xi, y - \eta) = \frac{1}{(2\pi)^2} \iint_{-\infty}^{+\infty} \frac{\tilde{\Pi}^*(p_x, p_y) e^{i[p_x(x - \xi) + p_y(y - \eta)]}}{\tilde{\Pi}^*(p_x, p_y)\tilde{\Pi}(p_x, p_y) + \alpha[1 + (p_x^2 + p_y^2)^2]^2} \, dp_x \, dp_y$$

where α is the regularisation parameter, an asterisk denotes complex conjugation and a tilde denotes Fourier transformation.

Thus, this state of atmospheric correction of satellite sensor data (which involves two-dimensional filtering) is in sharp contrast to pixel-by-pixel transformations (Equation 10.3) and consists of calculating a two-dimensional Fourier transformation and the convolution integral (Equation 10.16). The horizontal size of the window within which the filtering occurs is determined by the character of the processed scene, atmospheric conditions and the spatial resolution of the satellite sensor.

The transfer function for the satellite sensor (SS)–Earth's surface-atmosphere (SA) is represented as $\Pi(p_x, p_y)$. It is the product of

$$\Pi(p_x, p_y) = h_{SS}(p_x, p_y)\, h_{SA}(p_x, p_y)$$

where h_{SS} and h_{SA} are filters in the interval of spatial frequencies and can be approximated in the following way:

$$h_{SS}(p_x, p_y) = e^{-\beta_1 \sqrt{p_x^2 + p_y^2}} \quad h_{SA}(p_x, p_y) = e^{-1/\mu \sqrt{1-\mu^2}\,(p_x \cos\phi + p_y \sin\phi)}$$

$$\times (e^{-1/\mu} + S_{MS} e^{-\beta_2 \sqrt{p_x^2 + p_y^2}}) \quad (10.17)$$

Here S_{MS} is the rate of the multiple-scattering integral in the two-dimensional transfer equation. The β_2 values for the data in Figure 10.6 at $p_x = 2\cdot 5$ km^{-1} and $p_y = 0$ are given in Table 10.3. This shows that the problem of atmospherically correcting remote sensor data has two stages: pixel-by-pixel retrieval of mean surface albedo and transformation within a variably sized 'window', to retrieve the fluctuating component of albedo. The transformation functions for both of the processing stages are determined from Equations 10.13–10.17. The input data are the known survey angular coordinates, atmospheric turbidity parameters and some information on the Earth's surface (e.g. degree of non-orthotropicity, topography, spatial frequency, etc.), which are known *a priori*, are determined from measurements on test sites or are derived using the semivariograms of available remotely sensed imagery (Curran, 1988).

As an example, data from the Meteor MSU-M low-spatial-resolution $0\cdot 5$–$0\cdot 6$ μm channel were atmospherically corrected using data in Tables 10.2 and 10.3 (Kozoderov *et al.*, 1980b). A comparison of the uncorrected and corrected imagery showed differences similar to those observed by those who have atmospherically corrected Landsat MSS data (Rogers, 1973; Peacock, 1974), in that image contrast increased, classification accuracy increased, and so on. Thus, against the criterion traditionally used in assessing the value of atmospheric correction (i.e. the accuracy with which the data may be classified),

Table 10.3 *Values of the parameter for the exponential approximation of the spatial frequencies transfer function for the surface-atmosphere system for five mean albedos ($\bar{A}$).*

	$\bar{A}$ (per cent)				
τ	10	30	50	70	90
0·1	0·2231	0·2524	0·2870	0·3257	0·4064
0·5	0·3983	0·4634	0·5405	0·7246	0·9512

the application of the improved mathematical formalism discussed previously is a useful procedure and is now widely adopted in the USSR.

10.3. Concluding comments on the atmospheric correction of remotely sensed data

This chapter has shown that there is a possibility of accounting for atmospheric effects on remotely sensed data by using an improved mathematical formalism. A number of researchers (Rogers, 1973; Peacock, 1974; Switzer et al., 1981) have provided generalised optical characteristics (cross sections and indicatrices) of the atmosphere that will enable quantitative application of these formalisms. However, these characteristics can only be known approximately because of their natural spatial and temporal variability and because their retrieval methods are imperfect.

An important stage in assessing the accuracy of atmospheric correction is a comparison of atmospherically corrected data with near-surface spectral reflectance measurements. However, this is a complicated process, and one which has received little attention in the USSR. Frequently Soviet researchers in the field of remote sensing are faced with finding the solution to a given problem with a suboptimal data set. A reference image against which the effects of atmospheric correction could be compared would therefore be a valuable asset. However, acquiring such a reference image is a problem. A catalogue of the spectral characteristics of natural targets derived from satellite remote sensor data and a theory enabling these to be related to surface measurements may also be an aid and are available for certain test sites. In the future a better solution may be obtained from either the combined use of satellite sensor data and subsatellite measurements, both in regions of substantial atmospheric effect and in relatively transparent windows or by comparison with the established atmospheric transmission radiance models (e.g. LOWTRAN) that are well-used by researchers in the West (e.g. Kneizys et al., 1983).

11
Remote sensing of soil and crop state

Remote sensing research in the USSR has focused on the development of theory. This has been refined and validated by means of remotely sensed data recorded from aircraft and satellites. During the 1980s the potential for using this theory, in conjunction with imagery from Soviet satellites, has been evaluated. This chapter reports on two such evaluations.

11.1. Assessing the state of soils from aircraft and satellite sensor measurements

Soil state has an important influence on a variety of factors such as its erosivity, land use and land potential. The growing anthropogenic impacts on all aspects of the environment (Goudie, 1986), from local to global scales (NASA, 1988) cannot be measured adequately by traditional methods. The need for prompt and accurate information on soil properties for large areas of the USSR necessitated the development of reliable remote sensing techniques for assessing soil state. However, ground data are still required because successful solution of the inverse problems from remotely sensed data depends largely on the availability on an adequate *a priori* database for calibration.

Agricultural applications of information on soil state are many. For example, soil moisture content influences such factors as the optimal timing of tillage (Spoor, 1975) and soil humus content modulates the soil's stability, chemical composition and microclimate (Russell, 1973; Brady, 1984). As an example of soil state assessment from satellite remote sensor data, consideration will be given to the determination of the humus content of ploughed soils from medium spatial resolution Meteor sensor imagery. This work therefore builds on the work discussed in Chapter 5.

Many experiments have observed a close correlation between the spectral reflectance of a soil and its humus content (Pokrovsky, 1929; Obukhov and Orlov, 1964; Zyrin and Kuliev, 1967; Mikhailova, 1970; Tolchelnikov, 1974; Fedchenko 1982b, 1983; Kondratyev and Fedchenko, 1982e). Studies have shown that this can be reliably achieved using colorimetric techniques for quantitatively assessing soil colour. The most representative characteristic from which soil humus content can be quantitatively estimated is the colour coordinates (Fedchenko, 1982b, 1983; Kondratyev and Fedchenko, 1982f). As a basis for estimating soil humus content from space, initial measurements were made of soil spectral reflectance in the laboratory and field at both surface and aircraft altitudes.

Ploughed fields on either side of the Kremenchug reservoir in the Cherkassy and Kiev regions of the USSR (Figure 4.1) were selected as reference test sites. For each test site the humus content was estimated from SBCs derived using an airborne, fast-operating multichannel spectrometer (Koltsov, 1975; Fedchenko and Kondratyev, 1981). These data were acquired for 57 fields in August 1981, from an altitude of 100–150 m.

Colour coordinates were derived from the spectral-reflectance curves using Equation 2.9. The soil humus content was then estimated from the calculated values of the colour coordinates with the aid of a predrawn graph of the relationship (e.g. Figures 5.6, 5.9, 5.11 and 5.12). As discussed earlier, estimating soil humus content from colour coordinates in this fashion is advantageous because the entire soil spectrum is used.

The location of the test sites on the Meteor sensor imagery was a problem. In addition it was found to be impossible to reference satellite sensor measurements over individual fields precisely, because of the low spatial resolution (approximately 250 m) of the Meteor sensor. Therefore, the airborne sensor data were used to construct a training set from which the humus content in soils of a different territory could be estimated. The area around the Kakhovka reservoir, Ukraine, was selected for this porpuse.

Negatives of the satellite sensor imagery were scanned with a P-1000 photodensitometer (with a 100 μm aperture) and digitized to eight-bit precision. Points were selected on the imagery which coincided, at least as well as could be determined, with the aircraft measurements. These were described statistically and taken to form training sets for later classification (Borisoglebsky and Kozoderov, 1982).

Optical densities of the satellite sensor imagery were found to be closely correlated to the colour coordinates for the 57 points measured from the aircraft (Figure 11.1). However, a quantitative estimate of this correlation can only be made usefully when the aircraft and satellite sensor measurement are reliably referenced. The data used to construct Figure 11.1 show that the maximum error in determining the humus content of soils from the optical density of the satellite sensor imagery would not exceed 6 per cent. It was estimated that errors associated with incorrect referencing of the test sites constituted 1–2 per cent of the variability in optical density and that these errors increased the errors in humus content estimation by 2–3 per cent.

Classification of the satellite sensor imagery was error-prone because of the overlap in optical densities between the classes. For instance, the optical densities of turbid water and ploughed soils partially overlapped, as did ripe corn and

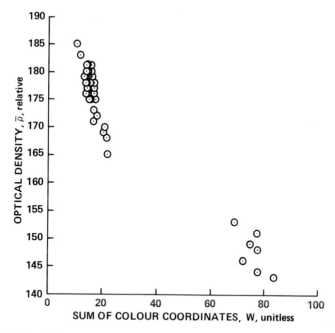

Figure 11.1 The relationship between colour coordinates derived from airborne spectrometer data and optical density derived from satellite sensor data for 57 fields.

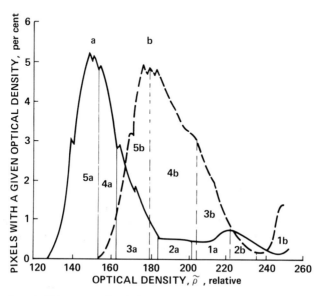

Figure 11.2 Smoothed histograms of optical densities for photographic negatives of a satellite sensor image. Figures correspond to the types of objects presented in Table 11.1, and a and b are two test sites.

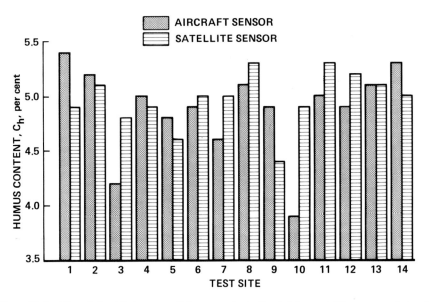

Figure 11.3 The relationship between soil humus content estimated from satellite sensor and aircraft sensor measurements of soil reflectance for blocks of fields in the Kherson region.

Table 11.1 Classification boundaries in graduations of optical density for satellite sensor negatives, where a and b are test sites (see Figure 11.2).

Target	Test sites	
	a	b
Water	>202	>235
Coniferous forest	184–202	222–235
Soils	163–184	204–222
Grass	153–163	178–204
Light-coloured features (sand, clouds, etc.)	<153	<178

clouds, and grass and deciduous canopies. The classification accuracy was therefore low. Classification boundaries are given in Table 11.1 with the class histograms in Figure 11.2. Note that the optical densities of the same objects on the imagery are shifted with respect to each other on different test sites. This bias is due to changes in the scan times of the P-1000 photodensitometer, but is not important because absolute values of the optical density are not required. The ploughed soil class can be further subdivided into two classes on the basis of differences in optical density. These differences are probably due to humus content because the type, moisture content and character of surface cultivation of these soils were similar at the time of measurement. In identifying these differences in optical density, attention should be given to the entire range of interrelated factors that characterise the soil state. However, most of these differences are usually accounted for by humus content in this region of the USSR. If other parameters of soil state have been recorded (e.g. erosivity, moisture content, iron content, etc.), then the optical density of the satellite sensor imagery may be related to them, even if the radiation–parameter relationship is non-causal.

A close correlation between humus contents estimated from airborne and spaceborne sensor data is apparent (Figure 11.3). Note that the airborne sensor measurements and laboratory analysis of the samples were undertaken at an earlier date (Fedchenko and Kondratyev, 1981). Although error was introduced because of changes in the conditions between measurements, the estimates of humus content were similar. The high coincidence of the results in Figure 11.3 can probably be accounted for by the low variability of the data in the training set (Figure 11.2) and a sufficiently reliable classification of the ploughed soils in the imagery (Table 11.1). A combined technique of using measurements from both satellite and aircraft to assess soil state is therefore promising.

11.2. Assessing the state of crops from satellite sensor measurements

The spatial variations of spectral brightness in agricultural areas, even when the crop is well established, is largely due to the heterogeneous state of the crops. This determines the spatial distribution of the reflected radiation field that is measured by satellite sensors. Assessing crop state from satellite sensor imagery in the USSR has, therefore, been based on analysing spatial variations.

Crop state is characterised by numerous parameters such as phytomass. Because spectral intervals can be selected where soil and vegetation display different SBCs (Figure 8.1), the seasonal increase of phytomass can be monitored by analysis of the changes in reflectance for the soil–vegetation system. Hence, the state of vegetation cover can be estimated from reflectance of the soil–vegetation system (Rachkulik and Sitnikova, 1981). An example of the relationships between crop state and reflectance for crops in the Kherson region is discussed below. First, the seasonal variations in the reflectance of the soil–crop system will be considered (Borisoglebsky and Kozoderov, 1982; Brisco et al., 1984).

Surface soil moisture content in the region shown in Figure 4.1 changes rapidly in early spring and late autumn. After snowmelt the surface soil layer changes from a sticky (overmoistened) state, to a soft-plastic state and then to a dry state. This drying of the surface layer usually takes 2–4 weeks, depending on the topography, geology and prevailing weather conditions. During this period soil reflectance changes markedly. An example for a chernozem soil typical of the Kherson region is shown in Figure 11.4 (Fedchenko and Kondratyev, 1981; Rachkulik and Sitnikova, 1981). In the summer the soil surface layer is mainly dry and soft-plastic, and its reflectance does not change substantially (Figure 11.4), although this is strongly modulated by precipitation.

Approximately 70 per cent of the land in the Kherson region is ploughed in spring and autumn. Consequently, variations in surface reflectance during these periods depend mainly on the moisture content of the soil surface layer (Figure 11.4). From the moment of surface drying (before the soft-plastic state) and resumed plant growth, surface reflectance changes are due mainly to increasing phytomass (Figure 11.4).

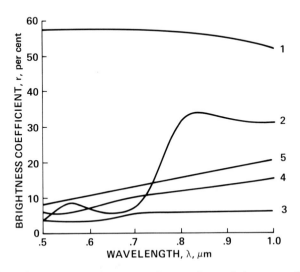

Figure 11.4 Spectral characteristics of (1) snow, (2) vegetation, and chernozem soil in a (3) sticky state, (4) soft-plastic state and (5) hard-plastic state, derived using a ground-based spectrometer.

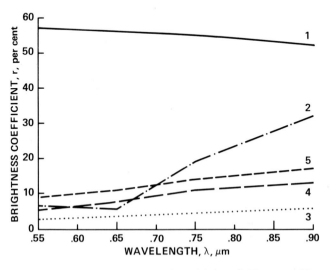

Figure 11.5 Atmospherically corrected spectral characteristics of (1) snow, (2) vegetation, and chernozem soil in a (3) sticky state, (4) soft-plastic state and (5) hard-plastic state, derived using four channels of the Meteor sensor.

Figure 11.5 gives the atmospherically and radiometrically corrected (Teillet, 1986) reflectance curves for snow, vegetation and soil in the wavebands of the Meteor sensor. This shows that the brightness coefficients of snow and vegetation differ substantially from that of soil, and that these differences are manifest in satellite sensor imagery. However, to analyse satellite sensor measurements quantitatively, the spectral sensitivity functions for the satellite radiometers must be known. Nevertheless, the data given in Figure 11.5 enable imagery to be classified, to a first approximation, using only image tone. Temporal variations in the ground coverage of green and senesced crops on ploughed soils are also apparent (Figure 11.6) and these data can be used in conjunction with satellite remote sensor imagery to assess the areas covered by crops in different states. Also in May, at least two classes—ploughed soils (dark) and crops (light)—have been identified because of different reflectances and image tones. However, the observed tones on any particular image vary with the percentage soil cover and the crop phenophases, etc.

To account for crop phenological factors when analysing the remotely sensed imagery, the basic crop development stages observed in the Kherson region are discussed, based on a study by Borisoglebsky and Kozoderov (1982). Relative to the other crops of the region, winter crops and multi-year grass are the first to complete their development cycle. When the winter wheat crop is earing, the early spring crops are still in the stalk-growth phase and late spring crops are generally in the plant-emergence and stalk-growth stages. A change in phytomass is associated with these phases of development. For instance, observations in the Kherson region on 28 May 1980 showed winter crops (earing phase) to be 70–80 cm high whereas early spring crops (stalk growth) were 25–50 cm high and late spring crops (plant-emergence/stalk-growth phase) were about 15 cm high. Grass, on the other hand, was green and at a lower height. Consequently,

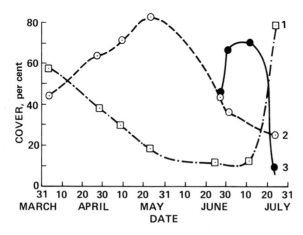

Figure 11.6 The seasonal dynamics of the percentage cover of (1) fallow soil, (2) green crops and (3) yellow crops for the Kherson region. Data from the Nizhnie Sergozy agricultural station for 1980.

the crops, because of their different phytomasses, soil cover, etc., displayed different brightnesses, so they could be identified on the imagery by their respective optical densities.

The crops differ in the timing of the onset of their phenological phases (Table 11.2). Also, the magnitude of the differences between onset dates for different crops changes throughout the cycle. For instance, the differences between wheat and spring barley in the initial period of their development are larger than during their ripening. Furthermore, the difference between the height of grass and that of the crops also declines, minimising their separability on the imagery. Late in the growing season the colours of the slow-ripening crops (maize and sunflower) differ greatly from the colour of the grass. Consequently, a crop calendar can be a valuable asset for crop classification (Brisco et al., 1984; Foody et al., 1989). Furthermore, the crop calendar can help to predict the most suitable periods for identifying and assessing crop state from remotely sensed data. In accordance with the data of Figures 11.5 and 11.6, Figure 11.7 presents the temporal dynamics of reflectance for winter and spring crops. Results of calculations based on the data of Figure 11.7 correspond to a description of the state of cereal crops given in Table 11.3. The percentage contribution by crops and soils to the total brightness of the soil–crop system was accounted for in the calculations. The

Table 11.2 *Differences in days between the beginning of the five major phenological stages of four agricultural crops, based on multi-year averages.*

Crop pairs	Phenological stages				
	Emergence	Tubing	Earing	Milk-ripeness	Wax-ripeness
Winter wheat–spring barley		19	6	5	4
Spring barley–maize	27		40		44
Sunflower–maize	15				13

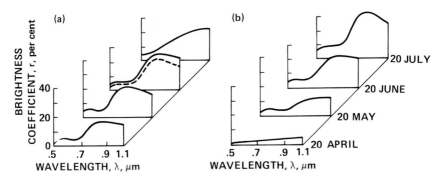

Figure 11.7 *Temporal change of reflectance for the soil–crop system: (a) winter crops and (b) spring crops. The winter crop in a poor state of health is represented by the broken line (Table 11.3).*

ratio of the area under crops to the total area under crops and soil was a weighting function.

Comparing the curves in Figure 11.7 shows that the brightness differences between the crops are small in visible wavelengths ($0\cdot5-0\cdot7\ \mu m$) and can be seen only in the transition from canopy to soil. In particular, there was little difference in brightness between the crops with low or high ground cover. However, there is a large increase in brightness contrast from visible to near-infrared wavelengths.

Once relationships between crop state and brightness have been formulated, the crop state may be estimated from remotely sensed data. However, the medium spatial resolution of the Meteor sensor (approximately 250 m) hinders per-field investigations, and groups of fields may have to be considered.

A crop classification was produced from the data on the assumption that a unique relationship exists between optical density and brightness. Optical densities corresponding to different crop states were classified for the region. The threshold values between optical densities were determined from characteristic break points of the histogram in Figure 11.8. This histogram is a distribution of ratios of the number of pixels with a given optical density to the total number of pixels in the area investigated. The peaks (nodes) and breaks of slope evident on the histogram indicate the presence of targets of different reflectances. From ground knowledge, it may be supposed that for the area investigated the targets

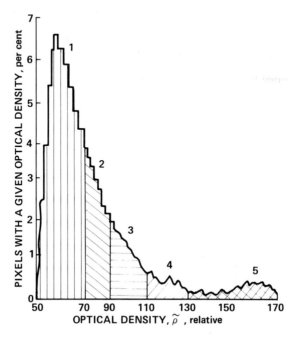

Figure 11.8 A histogram of optical densities of a Meteor sensor image of the Kerson region (28 May 1980): (1) winter crops, (2) early spring crops, (3) late spring crops, (4) soil and (5) reservoirs.

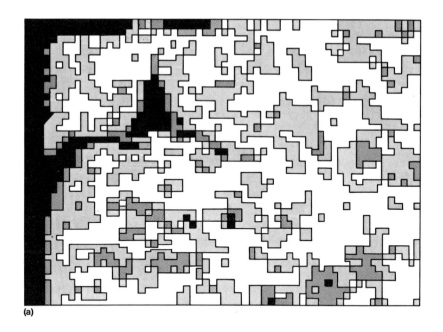

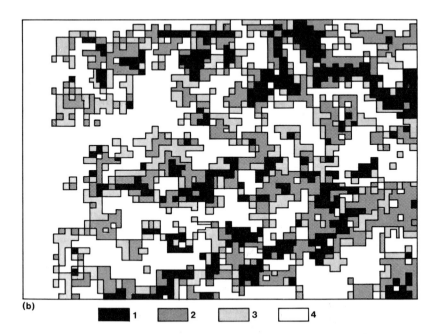

Figure 11.9 Two thematic maps produced from a Meteor sensor image of 28 May 1980 for the Kherson region. (a) Land cover: (1) reservoirs, (2) ploughed soils, (3) spring crops and (4) winter crops (Table 11.4). (b) State of winter crops: (1) above average, (2) average, (3) below average and (4) other (Table 11.5).

shown on the histogram (Figure 11.8) are water, ploughed soil and intermediate areas with variable proportions of soil and crops.

A survey on 28 May 1980 showed that winter crops, early spring crops and late spring crops differed in height. This knowledge and the differences in the spectral brightnesses of the crops allowed five classes to be selected on the break points in Figure 11.8; winter crops, ploughed soils, reservoirs and early- and late-spring crops. The threshold levels from the histogram were then applied to the image. Thus, the image was classified into four classes (Figure 11.9a) on the basis of optical densities; using sliding averages of the data of Figure 11.8 (Kondratyev and Fedchenko, 1982a).

A comparison of ground data on the spatial disposition of crop types in the Kherson region (obtained *in situ* by the Moscow State University (MSU) expedition) showed considerable agreement with results from the classified remotely sensed data (Table 11.4). The accuracy of the area estimations from the satellite remote sensor data depends largely on the correct selection of the threshold optical densities based on *a priori* information and, in this example, it comprised *a priori* information on the phase of crop development.

Areas classified as belonging to the same crop type displayed variations in

Table 11.3 *The phenological stages of the cereals for which reflectance is reported in Figure 11.7. The vegetation cover of the winter crop in a poor state of health is given in per cent in parentheses.*

	Winter crop			Spring crop		
Date	Phenological stage	Colour	Vegetation cover	Phenological stage	Colour	Vegetation cover
20 April	Tubing	Light green	10	Sprouting	Dark	0
20 May	Earing	Green	50	Bushing	Light green	30
20 June	Blossoming	Dark green	90 (50)	Earing	Green	60
20 July	Stubble	Yellow	40	Ripening	Yellow	90

Table 11.4 *Five classes of land cover for a portion of the Meteor sensor image of the Kherson region for 28 May 1980 (Figure 11.9a).*

	Area (per cent)	
Land cover	From optical densities of Meteor sensor image	From MSU expedition
Reservoirs	8	8
Ploughed soils	12	15
Late-spring crops	16	20
Early-spring crops	20	11
Winter crops	44	46

Table 11.5 *The state of winter crops for a portion of the Meteor sensor image of the Kherson region for 28 May 1980 (Figure 11.9b).*

State of winter crops	Area (per cent)	
	From optical densities of Meteor sensor image	From MSU expedition
Above average (High percentage cover)	22	24
Average (Medium percentage cover)	42	51
Below average (Low percentage cover)	36	25

optical density, variations which are due largely to crop state. For instance, on positive images the optical density decreased with increasing soil cover. Thus, areas of different crop state can be delimited and quantitative relationships between optical density and crop state (phytomass, etc.) can be sought. The state of winter wheat in part of the Kherson region is shown in Figure 11.9b. The winter wheat had been subdivided into three class states: high, medium and low percentage cover. Threshold values were determined and the image was classified. It was found that there was a degree of correspondence between the classified data and the *in situ* crop state measurements acquired by the MSU expedition (Table 11.5). Consequently, the technique may be applied, with reasonable accuracy, to assessing crop state over large areas.

It must be noted, however, that the choice of threshold levels is subjective. These levels, as well as the interrelationships between optical density levels and quantitative indices of state, must be determined on the basis of models of the radiation/soil–crop system relationship and *a priori* information. Thus, although it is difficult to accomplish a brightness correction of the satellite sensor imagery and subjective decisions are required there is evidence that crop state can be assessed from imagery with a medium spatial resolution. Future research in the USSR will concentrate on the use of the high spatial and spectral-resolution imagery and the utilisation of temporal reflectance profiles (Kondratyev *et al.*, 1986b).

11.3. Concluding comments on the remote sensing of soil and crop state

The two experiments reported in this chapter highlight the problems of trying to relate point measurements for a local area on the ground to satellite sensor data for a region. This fundamental problem of scaling-up from local to regional estimates of soil and crop state is becoming a key research area in the USSR. The results of this Soviet research are likely to command considerable attention as the current global change initiatives (NASA, 1988) come to fruition in the 1990s.

Bibliography

Adams, J. B., Smith, M. O. and Johnson, P. E., 1986, Spectral mixture modelling: a new analysis of rock and soil types at the Viking Lander 1 site. *Journal of Geophysical Research*, **91**, 8098–8112.
Ahern, F. J., Brown, R. J., Cihlar, J., Gauthier, R., Murphy, J., Neville, R. A. and Teillet, P. M., 1987, Radiometric correction of visible and infrared remote sensing data at the Canada Centre for Remote Sensing. *International Journal of Remote Sensing*, **8**, 1349–1376.
Akhmanov, S. A., Dyakov, Yu. E. and Chirkin, A. S., 1981, *Introduction to Statistical Radiophysics and Optics* (Moscow: Nauka).
Antokolsky, M. L., 1948, Reflection of waves from a rough absolutely reflecting surface. *Doklady USSR Academy of Sciences*, **62**, 203–206.
Asmus, V. V., Spiridonov, Yu. G. and Tishenko, A. P., 1980, Practical aspects of radiometric correction of multizonal videoinformation. *Studies of the Earth from Space*, **1**, 59–68.
Barahanenkov, Yu. N., 1973, On wave corrections to a transfer equation for backscattering. *Izvestia Vuzov Radiophysics*, **16**, 88–95.
Baslavskaya, S. S. and Trubetskova, O. N., 1964, *Practicum on Plant Physiology* (Moscow: Moscow State University Press).
Bass, F. G. and Fuks, I. M., 1972, *Wave Scattering on a Statistically Rough Surface* (Moscow: Nauka).
Beckett, P. H. T. and Webster, R., 1971, Soil variability: a review. *Soils and Fertilizers*, **34**, 1–15.
Benedict, H. M. and Swidler, R., 1961, Non-destructive method for estimating chlorophyll content of leaves. *Science*, **133**, 2015–2016.
Borisoglebsky, G. I. and Kozoderov, V. V., 1982, An assessment of the state of crops from satellite data. *Meteorology and Hydrology*, **3**, 94–97.
Born, M. and Wolf, E., 1973, *Fundamentals of Optics* (Moscow: Nauka).
Bouma, P. J., 1971, *Physical Aspects of Colour*, 2nd edition (London: Macmillan).
Bowers, S. A. and Hanks, R. J., 1965, Reflection of radiant energy from soil. *Soil Science*, **100**, 130–138.
Brady, N. C., 1984, *The Nature and Properties of Soils*, 9th edition (New York: Macmillan).
Brandt, A. B. and Tageeva, S. V., 1967, *Optical Parameters of Vegetation* (Moscow: Nauka).
Brekhovskikh, L. M., 1952a, Wave diffraction on a rough surface. General theory. *Journal of Experimental and Theoretical Physics*, **23**, 275–288.
Brekhovskikh, L. M., 1952b, Wave diffraction on a rough surface. Appendices to general theory. *Journal of Experimental and Theoretical Physics*, **23**, 289–304.
Bridges, E. M., 1978, *World Soils*, 2nd edition (Cambridge: Cambridge University Press).
Briggs, D. J. and Courtney, F. M., 1985, *Agriculture and Environment* (London: Longman).

Brisco, B., Ulaby, F. T. and Protz, R., 1984, Improving crop classification through attention to the timing of airborne radar acquisitions. *Photogrammetric Engineering and Remote Sensing*, **50**, 739–745.

Chamberlin, G. J., 1951, *The C.I.E. International Colour System Explained* (Salisbury: The Tintometer).

Chaume, D. and Phu, T. N., 1980, Interactive processing of Landsat image for morphopedological studies. *Machine Processing of Remotely Sensed Data Symposium*, West Lafayette, Laboratory for the Applications of Remote Sensing, pp 195–204.

Colwell, J. E., 1974, Grass canopy bidirectional spectral reflectance. *Proceedings, 9th International Symposium on Remote Sensing of Environment* (Ann Arbor: University of Michigan), pp 1061–1085.

Committee on Colorimetry, 1963, *The Science of Color* (Washington, DC: Optical Society of America).

Condit, H. R., 1970, The spectral reflectance of American soils. *Photogrammetric Engineering*, **36**, 955–966.

Curran, P. J., 1980, Relative reflectance data from preprocessed multispectral photography. *International Journal of Remote Sensing*, **1**, 77–83.

Curran, P. J., 1985a, *Principles of Remote Sensing* (London: Longman Scientific and Technical).

Curran, P. J., 1985b, Aerial photography for the assessment of crop condition: a review. *Applied Geography*, **5**, 347–360.

Curran, P. J., 1985c, The A. I. Voeikov Main Geophysical Observatory, Leningrad, USSR. *Weather*, **40**, 185–186.

Curran, P. J., 1988, The semi-variogram in remote sensing: an introduction. *Remote Sensing of Environment*, **24**, 493–507.

Curran, P. J. and Hay, A. M., 1986, The importance of measurement error for certain procedures in remote sensing at optical wavelengths. *Photogrammetric Engineering and Remote Sensing*, **52**, 229–241.

Curran, P. J. and Milton, E. J., 1983, The relationship between the chlorophyll concentration, LAI and reflectance of a simple vegetation canopy. *International Journal of Remote Sensing*, **4**, 247–256.

Curtis, L. F., Courtney, F. M. and Trudgill, S., 1976, *Soils in the British Isles* (London: Longman).

Dave, J. V., 1975a, A direct solution of the spherical harmonics approximation to radiative transfer equation for an arbitrary solar elevation. Part 1—Theory. *Journal of Atmospheric Science*, **32**, 790–798.

Dave, J. V., 1975b, A direct solution to the spherical harmonics approximation to radiative transfer equation for an arbitrary solar elevation. Part 2—Results. *Journal of Atmospheric Science*, **32**, 1463–1474.

Dave, J. V., 1980, Effect of atmospheric conditions on remote sensing of a surface nonhomogeneity. *Photogrammetric Engineering and Remote Sensing*, **46**, 1173–1180.

Dave, J. V. and Canosa, J. A., 1974, A direct solution of the radiative transfer equation: application to the atmospheric models with arbitrary vertical nonhomogeneities. *Journal of Atmospheric Science*, **31**, 1089–1101.

Dawson, J. H. and Holstun, J. T., 1970, *Estimating Losses from Weeds in Crops*, Crop Loss Assessment Methods (Rome: Food and Agriculture Organisation).

Deering, D. W. and Eck, T. F., 1987, Atmospheric optical depth effects on angular anisotropy of plant canopy reflectance. *International Journal of Remote Sensing*, **8**, 893–916.

Dergacheva, M. I., 1972, Optical properties of the system of humus substances of forested soils of the Urals and Zauralye. *Proceedings of the Institute of Ecology, Plants and Animals*, **85**, 146–155.

Dergacheva, M. I. and Kuz'mina, E. F., 1978, Spectral characteristics of humus acids of

Bibliography

salt soils of Altai Krai. In *Soil Properties in the Taiga and Forest-Steppe Zones of Siberia* (Novosibirsk: Nauka), pp 88–96.

Drury, S. A., 1987, *Image Interpretation in Geology* (London: Allen and Unwin).

Duggin, M. J., 1974, On the natural limitations of target differentiation by means of spectral discrimination techniques. *Proceedings, 9th International Symposium on Remote Sensing of Environment* (Ann Arbor: University of Michigan), pp 499–514.

Dzhad, D. and Vyshetski, G., 1978, *Colour in Science and Technology* (Moscow: Mir).

Eaton, F. D. and Dirmhirn, I., 1979, Reflected irradiance indicatrices of natural surfaces and their effect on albedo. *Applied Optics*, **18**, 994–1008.

Escadafal, R., Girard, M. and Coorault, D., 1989, Munsel soil color and soil reflectance in the visible spectral bands of Landsat MSS and TM data. *Remote Sensing of Environment*, **27**, 37–46.

Evans, R., 1972, Air photographs for soil survey in lowland England: soil patterns. *Photogrammetric Record*, **7**, 302–322.

Fedchenko, P. P., 1981, On some factors affecting soil spectral reflectance. *Trudy VHIISKhM*, **5**, 42–51.

Fedchenko, P. P., 1982a, A determination of the humus content in soils from their colour. *Soviet Soil Science*, **10**, 138–142.

Fedchenko, P. P., 1982b, Possibilities to estimate humus in soils from spectral measurements. *Studies of the Earth from Space*, **5**, 72–79.

Fedchenko, P. P., 1982c, *A Remote Sensing Technique to Assess the Green Crop Canopy Mass.* Invention Bulletin, A.C. No. 969204 (Moscow: USSR Technical Press).

Fedchenko, P. P., 1983, An estimation of humus in soils from spectral measurements. *Geography and Earth's Resources*, **2**, 68–78.

Fedchenko, P. P. and Kondratyev, K. Ya., 1981, *Spectral Reflectances of Some Soils* (Leningrad: Gidrometeoizdat).

Feldbaum, A. A. and Butkovsky, A. G., 1971, *Techniques in the Theory of Automatic Operation* (Moscow: Nauka).

Finkelberg, V. M., 1967, Wave propagation in a random medium. Method of correlative groups. *Journal of Experimental and Theoretical Physics*, **53**, 401–416.

Fisiunov, A. V., 1973, *Weeds and Measures Against Them* (Moscow: Znanie).

Fisiunov, A. V. and Matiukha, L. A., 1972, Special features of the technique for mapping crop and soil weediness. *Bulletin of the All-Union Institute for Maize Research*, **5–6**, 51–54.

Foody, G. M. and Wood, T. F., 1987, The use of Landsat TM data in a GIS for environmental monitoring. In *Advances in Digital Image Processing* (Nottingham: Remote Sensing Society) pp 434–439.

Foody, G. M., Curran, P. J., Groom, G. B. and Munro, D. C., 1989, Multi-temporal airborne synthetic aperture radar data for crop classification. *Geocarto International*, **4**, 19–29.

Forster, B. C., 1984, Derivation of atmospheric correction procedures for Landsat MSS with particular reference to urban data. *International Journal of Remote Sensing*, **5**, 799–817.

Fryer, J. D. and Evans, S. A., 1968, *Weed Control Handbook*, Volume 1, *Principles*, 5th edition (Oxford: Blackwell).

Gausman, H. W. and Allen, W. A., 1973, Optical parameters of leaves of 30 plant species. *Plant Physiology*, **52**, 57–62.

Gazarian, Yu.L., 1969, On a 1-D problem of wave propagation in a medium with random inhomogeneities. *Journal of Experimental and Theoretical Physics*, **56**, 1856–1871.

Goncharsky, A. V., Yagola, A. G. and Leonov, A. S., 1971, On solution of Fredholm 2-D integral equations of the first kind. *Journal of Experimental and Theoretical Physics*, **11**, 1296–1301.

Goudie, A., 1986, *The Human Impact on the Natural Environment*, 2nd edition (Oxford: Blackwell).
Griggs, M., 1975, Measurements of atmospheric aerosol optical thickness over water using ERTS-1 data. *Journal of the Air Pollution Control Association*, **25**, 622–626.
Gurevich, M. M., 1950, *Colour and its Measurement* (Moscow: AN SSSR).
Gurevich, M. M., 1968, *Introduction to Photometry* (Leningrad: Energy).
Guyot, G., Guyon, D. and Riom, J., 1989, Factors affecting the spectral response of forest canopies: A review. *Geocarto International*, **4**, 3–18.
Horler, D. N. H., Dockray, M. and Barber, J., 1983, The red edge of plant leaf reflectance. *International Journal of Remote Sensing*, **4**, 273–288.
Horwitz, W., 1970, *Official Methods of Analysis*, 11th edition (Washington, DC: Association of American Chemists).
Hunt, R. W. G., 1987, *Measuring Colour* (Chichester: Ellis Horwood).
Hunter, R. S. and Harold, R. W., 1987, *The Measurements of Appearance*, 2nd edition (New York: Wiley).
Isakovich, M. A., 1952, Wave scattering from a statistically rough surface. *Journal of Experimental and Theoretical Physics*, **23**, 305–314.
Isimaru, A., 1981a, *Propagation and Scattering of Waves in Randomly Inhomogeneous Media*, Volume I, *Single Scattering and the Transfer Theory* (Moscow: Mir).
Isimaru, A., 1981b, *Propagation and Scattering of Waves in Randomly Inhomogeneous Media*, Volume II, *Multiple Scattering, Turbulence, Rough Surfaces and Remote Sensing* (Moscow: Mir).
Johanssen, C. J. and Dacosta, L. H., 1980, Using soil color reflectance in predicting soil properties. In *Machine Processing of Remotely Sensed Data Symposium* (West Lafayette: Laboratory for the Applications of Remote Sensing), pp 233–237.
Judd, D. B. and Wyszecki, G., 1975, *Color in Business and Industry*, 3rd edition (New York: Wiley).
Karmanov, I. I., 1974, *Spectral Reflectance and Colour of Soils as Indicators of their Properties* (Moscow: Kolos).
Kastrov, B. G., 1955, On the diurnal change of surface albedo. *Trudy CAO*, **14**, 12–22.
Kauli, J., 1979, *Physics of Diffraction* (Moscow: Mir).
Kelly, K. L. and Judd, D. B., 1976, *Color: Universal Language and Dictionary of Names* (Washington, DC: National Bureau of Standards (USA)), Special Publication 440.
Khabibrakhmanov, Kh., 1974, *Studies of Field Weediness to Develop Measures Against Weeds* (Leningrad: Kazan).
Kharin, N. G., 1975, *Remote Sensing of Vegetation* (Moscow: Nauka).
Kimes, D. S. and Kirchner, J. A., 1982, Irradiance measurement errors due to the assumption of a Lambertian reference panel. *Remote Sensing of Environment*, **12**, 141–149.
Kimes, D. S., Sellers, P. J. and Diner, D. J., 1987, Extraction of spectral hemispherical reflectance (albedo) of surfaces from nadir and directional reflectance data. *International Journal of Remote Sensing*, **8**, 1727–1746.
Kizel, V. A., 1973, *Reflection of Light* (Moscow: Nauka).
Kliatskin, V. I., 1980, *Stochastic Equation for Waves in Randomly Inhomogeneous Media* (Moscow: Nauka).
Kneizys, F. X., Shettle, E. P., Gallery, W. O., Chetwynd, J. H., Albrell, L. W., Selby, J. E. A., Clough, S. A. and Fenn, R. W., 1983, *Atmospheric Transmission/Radiance: Computer Code Lowtran 6*, Environmental Paper Number 846, Air Force Geophysical Laboratory.
Koltsov, V. V., 1975, A field fast-operating spectrometer with discrete scanning of the spectrum. *Meteorology and Hydrology*, **8**, 100–103.
Kondratyev, K. Ya. and Fedchenko, P. P., 1979, Spectral reflectance of some weeds. *Doklady USSR Academy of Sciences*, **248**, 1318–1320.

Kondratyev, K. Ya. and Fedchenko, P. P., 1980a, The diurnal change of spectral reflectances of soils and vegetation. *Studies of the Earth from Space*, **1**, 40–47.

Kondratyev, K. Ya. and Fedchenko, P. P., 1980b, Possibilities of using the soil reflectance spectra to study their properties. *Studies of the Earth from Space*, **1**, 114–124.

Kondratyev, K. Ya. and Fedchenko, P. P., 1980c, The effect of processing on the spectral reflectance of soils. *Soviet Soil Science*, **8**, 47–52.

Kondratyev, K. Ya. and Fedchenko, P. P., 1980d, An experience of recognition of some crops from their reflectance spectra. *Studies of the Earth from Space*, **1**, 50–55.

Kondratyev, K. Ya. and Fedchenko, P. P., 1980e, Recognition of soils by their reflection spectra. In *Proceedings, 14th International Symposium on Remote Sensing of Environment* (Ann Arbor: University of Michigan), pp 409–413.

Kondratyev, K. Ya. and Fedchenko, P. P., 1981a, The use of remote sensing techniques to study soils. *Geography and Earth's Resources*, **3**, 83–90.

Kondratyev, K. Ya. and Fedchenko, P. P., 1981b, Recognition of some crops from their reflection spectra. *Advances in Space Research*, **1**, 87–88.

Kondratyev, K. Ya. and Fedchenko, P. P., 1982a, A determination of corn weediness from spectral measurements. *Studies of the Earth from Space*, **3**, 59–68.

Kondratyev, K. Ya. and Fedchenko, P. P., 1982b, *Spectral Reflectance and Recognition of Vegetation Cover* (Leningrad: Gidrometeoizdat).

Kondratyev, K. Ya. and Fedchenko, P. P., 1982c, Remote sensing of areas under damaged and dead winter crops. *Meteorology and Hydrology*, **8**, 102–108.

Kondratyev, K. Ya. and Fedchenko, P. P., 1982d, *The Use of Remote Sensing Techniques to Study the State of Crops. Overview* (Obninsk: Hydrometeorology).

Kondratyev, K. Ya. and Fedchenko, P. P., 1982e, On studies of humus in soils by their colour. *Doklady USSR Academy of Sciences*, **263**, 988–999.

Kondratyev, K. Ya. and Fedchenko, P. P., 1982f, On measurements of soil colour. *Soviet Soil Science*, **10**, 109–114.

Kondratyev, K. Ya. and Fedchenko, P. P., 1982g, A technique for determining the areas under dead winter crops. *Doklady USSR Academy of Sciences*, **256**, 1512–1514.

Kondratyev, K. Ya. and Fedchenko, P. P., 1982h, An experience in using reflectance spectra to recognise crops. *Studies of the Earth from Space*, **3**, 48–51.

Kondratyev, K. Ya. and Smoktiy, O. I., 1972, On the determination of the transfer function in spectrophotometry of the planetary surface from space. *Doklady USSR Academy of Sciences*, **206**, 1102–1105.

Kondratyev, K. Ya. and Smoktiy, O. I., 1973, On the determination of the spectral transfer function for the brightness of natural formations and their contrasts in spectrophotometry of the atmosphere–surface system from space. *Trudy MGO*, **295**, 24–50.

Kondratyev, K. Ya., Badinov, I. Ya., Ivlev, L. S. and Nikolsky, G. A., 1969, Aerosol structure of the troposphere and stratosphere. *Physics of the Atmosphere and Ocean*, **5**, 480–493.

Kondratyev, K. Ya., Korzov, V. I. and Ter-Markaryants, N. E., 1974, Optical inhomogeneity of the urban surface from aircraft measurements during the CAENEX-72 expedition. *Trudy MGO*, **331**, 41–49.

Kondratyev, K. Ya., Buznikov, A. A., Vasilyer, O. B. and Smoktiy, O. I., 1975, The effect of the atmosphere on the spectral brightnesses and contrasts of natural formations when spectrometering the Earth from Space. *Physics of the Atmosphere and Ocean*, **11**, 348–361.

Kondratyev, K. Ya., Korzov, V. I. and Ter-Markaryants, N. E., 1976, Reflection characteristics of the surface-atmosphere system in conditions of large dust loading. *Trudy MGO*, **370**, 119–128.

Kondratyev, K. Ya., Fedchenko, P. P. and Barmina, Yu. M., 1982a, An experience in determining the chlorophyll content in the leaves of plants from colour coordinates. *Doklady USSR Academy of Sciences*, **262**, 1022–1024.

Kondratyev, K. Ya., Kozoderov, V. V. and Fedchenko, P. P., 1982b, Possibilities of determining the chlorophyll content in plants from their reflectance spectra. *Studies of the Earth from Space*, **3**, 63–68.

Kondratyev, K. Ya., Kozoderov, V. V., Fedchenko, P. P., and Barmina, Yu. M., 1982c, On the determination of the chlorophyll content in the leaves of plants from data of spectral measurements. *Doklady USSR Academy of Sciences*, **256**, 1508–1510.

Kondratyev, K. Ya., Kozoderov, V. V. and Fedchenko, P. P., 1986a, *Aerospace Studies of Soils and Vegetation* (Leningrad: Gidrometeoizdat).

Kondratyev, K. Ya., Kozoderov, V. V. and Fedchenko, P. P., 1986b, Remote sensing of the state of crops and soils. *International Journal of Remote Sensing*, **7**, 1213–1235.

Konova, M. M., 1963, *The Organic Matter of Soil* (Moscow: AN SSSR).

Korzov, V. I., 1973, Aircraft spectral measurements of the angular distribution of shortwave radiation reflected from stratus clouds. *Trudy MGO*, **317**, 17–27.

Korzov, V. I. and Krasilshchikov, L. B., 1974, A technique for and some results from aircraft measurements of the angular and spectral distribution of reflected shortwave radiation. In *Radiative Processes in the Atmosphere and on the Surface*, Proceedings of the All-Union Symposium on Actimetry (Leningrad: Gidrometeoizdat), pp 57–61.

Korzov, V. I. and Ter-Markaryants, N. E., 1976, Aircraft measurements of the angular and spectral characteristics of solar radiation reflection under the GATE programme. In *Proccedings of the GATE International Expedition, TROPEX-74* (Leningrad: Gidrometeoizdat), pp 556–563.

Kozoderov, V. V., 1977, An assessment of the atmospheric distortion in decoding of natural formations from space. In *Aerospace Studies of the Earth. Techniques for Processing Videoinformation with the use of Computers* (Moscow: Nauka), pp 24–35.

Kozoderov, V. V., 1982, The use of electromagnetic field equations to describe the interaction between radiation and natural formations. *Studies of the Earth from Space*, **3**, 83–87.

Kozoderov, V. V. and Mishin, I. V., 1980, On calculations of the spatial frequency characteristic of the surface-atmosphere system using the technique of spherical harmonics. *Trudy GosNITsIPR*, **8**, 64–74.

Kozoderov, V. V. and Shulgina, N. B., 1980, On application of electromagnetic field equations to the problem of remote sensing data interpretation. In *Abstract of Papers from the XI All-Union Symposium on Actinometry*, Tallin, pp 18–21.

Kozoderov, V. V., Sazhina, L. M. and Shulgina, N. B., 1980a, A mathematical model for transformation in the atmosphere of radiative signatures of natural objects. *Trudy GosNITsIPR*, **8**, 35–50.

Kozoderov, V. V., Liubovny, N. D. and Tishenko, A. P., 1980b, Radiometric correction of surface images from space. *Trudy GosNITsIPR*, **8**, 83–87.

Küchler, A. W. and Zonneveld, I. S., 1988, *Vegetation Mapping* (The Hague: Kluwer Academic).

Kulebakin, V. S., 1926, On reflection of the light from soil covers. *Proceedings of the State Electrotechnical Experimental Institute*, **17**, 5–28.

Latz, K., Weismiller, R. A. and Van Scoyoc, G. E., 1981, *A Study of the Spectral Reflectance of Selected Eroded Soils of Indiana in Relationship to their Chemical and Physical Properties* (West Lafayette: Laboratory for the Applications of Remote Sensing), Technical Report.

Le Grand, Y., 1968, *Light and Colour Vision*, 2nd edition (London: Chapman and Hall).

Leshkevich, G. A., 1988, Non-Lambertian reference panel effect on spectral reflectance measurements of freshwater ice. *International Journal of Remote Sensing*, **9**, 825–832.

Li, R. Y., Ulaby, F. T. and Eyton, J. R., 1980, Crop classification with a Landsat/radar sensor combination. In *Machine Processing of Remotely Sensed Data Symposium* (West Lafayette: Laboratory for the Applications of Remote Sensing), pp 78–87.

Liberstein, I. I., 1973, On recording the fields weediness. *Agriculture in Moldavia*, **6**, 33–34.

Lillesand, T. M. and Kiefer, R. W., 1987, *Remote Sensing and Image Interpretation*, 2nd edition (New York: Wiley).
MacAdam, D. L., 1985, *Color Measurement*, 2nd edition (New York: Springer-Verlag).
Maltsev, A. I., 1933, *Weeds in the USSR Territory* (Moscow and Leningrad: State-Farm and Collective-Farm Literature).
Mani, V. S., Gautan, K. G. and Chakraborty, T. K., 1968, Losses in crop yield in India due to weed growth. *Test Articles and News Summaries*, **14**, 142–158.
Markov, M. V., 1970, *Weeds in the Fields and Techniques for their Study* (Leningrad: Kazan).
Mather, P. M., 1987, *Computer Processing of Remotely Sensed Images* (Chichester: Wiley).
Merik, B., 1968, *Light and Color Measurements of Small Light Sources* (New York: General Electric Company).
Mikhailova, N. A., 1970, Correlation between humus content in soils and their spectral reflectance. *Soil and Agrochemical Studies in the Far East*, **19**, 52–59.
Milton, E. J., 1980, A portable multiband radiometer for ground data in remote sensing. *International Journal of Remote Sensing*, **1**, 153–165.
Milton, E. J., 1987, Principles of field spectroscopy. *International Journal of Remote Sensing*, **8**, 1807–1827.
Milton, E. J., 1989, On the suitability of Kodak Neutral Test Cards as reflectance standards. *International Journal of Remote Sensing*, **10**, 1041–1047.
Mishin, I. V. and Sushkeich, T. A., 1980, The optical spatial-frequency characteristic and its application. *Studies of the Earth from Space*, **4**, 69–80.
Moiseichik, V. A., 1978, *Long-range Agrometeorological Forecasts of Winter Crops Wintering Over the Territories of Regions, Republics and in the USSR, on the Whole* (Leningrad: Gidrometeoizdat).
Mueller-Dumbois, D. and Ellenberg, H., 1974, *Aims and Methods of Vegetation Ecology* (New York: Wiley).
Myers, V. I., 1983, Remote sensing applications in agriculture. In *Manual of Remote Sensing*, edited by R. N. Colwell (Falls Church, Virginia: American Society of Photogrammetry), pp 2111–2225.
NASA, 1988, *Earth System Science: A Closer View* (Washington, DC: National Aeronautics and Space Administration).
Neema, D. L., Shah, A. and Patel, A. N., 1987, A statistical optical model for light reflection and penetration through sand. *International Journal of Remote Sensing*, **8**, 1209–1217.
Nichiporovich, A. A., 1955, *The Light and Carbon Feeding of Plants (Photosynthesis)* (Moscow: AN SSSR).
Nikolsky, V. V., 1978, *Electrodynamics and Radiowave Propagation* (Moscow: Nauka).
Obukhov, A. I. and Orlov, D. S., 1964, Spectral reflectance of major types of soils and possibility of using diffuse reflection in soil studies. *Soviet Soil Science*, **1**, 174–184.
Ogorodnikov, B. I. (editor), 1971, *Manual on Visual Aircraft Agrometeorological Studies* (Moscow: Gidrometeoizdat).
Orlov, D. S., 1960, A technique to study optical properties of humus substances. *Science Papers of High School Biology*, **1**, 204–207.
Orlov, D. S., 1966a, Quantitative laws of reflection of light by soils. The effect of particle size (aggregates) on reflectance. *Science Papers of High School Biology*, **4**, 206–210.
Orlov, D. S., 1966b, Spectrophotometrical analysis of humus substances. *Soviet Soil Science*, **11**, 84–95.
Orlov, D. S., 1969, Quantitative laws of reflection of light by soils IV. Varying indices and the effect of humus substances. *Science Papers of High School Biology*, **8**, 131–134.
Orlov, D. S. and Grishina, L. A., 1981, *Practicum on Humus Chemistry* (Moscow: Moscow State University Press)
Orlov, D. S. and Proshina, N. V., 1975, Quantitative laws of reflection of light by soils. VI. The variative-statistical characteristic of soils at the Zvenigorod biological station of Moscow University. *Science Papers of High School Biology*, **17**, 111–114.

Orlov, D. S., Glebova, G. I. and Midakova, K. E., 1966, An analysis of the vertical distribution in soils of iron oxide compounds and humus from spectral brightness curves. *Science Papers of High School Biology*, **4**, 217–222.

Orlov, D. S., Bildebaeva, R. M. and Sadovnikov, Yu. N., 1976, Quantitative laws of reflection of light by soils. VII. Spectral reflectance of major soils in Kazakhstan. *Science Papers of High School Biology*, **20**, 109–113.

Orlov, D. S., Obukhov, A. I., Vedina, O. T. and Sadovnikov, Yu. B., 1978, Quantitative laws of reflection light by soils. Sub-tropical soils of Western Georgia. *Scientific Papers of High School Biology*, **22**, 127–132.

Panfilov, A. S., 1976, Presentation of reflectances for natural materials. *Studies of the Earth from Space*, **2**, 316–318.

Peacock, K., 1974, Ground-based determination of atmospheric radiance for correction of ERTS-1 data. *Applied Optics*, **13**, 2741–2742.

Philipson, W. R., Gordon, D. K., Philpot, W. D. and Duggin, M. J., 1989. Field reflectance calibration with grey standard reflectors. *International Journal of Remote Sensing*, **10**, 1035–1039.

Pickup, G. and Chewings, V. H., 1988, Forecasting patterns of soil erosion in arid lands from Landsat MSS data. *International Journal of Remote Sensing*, **9**, 69–84.

Pickup, G. and Nelson, D. J., 1984, Use of Landsat radiance parameters to distinguish soil erosion, stability and deposition in arid central Australia. *Remote Sensing of Environment*, **16**, 195–209.

Pokrovsky, G. I., 1929, On optical studies of humus in soils. *Soviet Soil Science*, **1–2**, 124–130.

Rachkulik, V. I. and Sitnikova, M. V., 1981, *Reflectances and the State of Vegetation Cover* (Leningrad: Gidrometeoizdat).

Robinove, C. J., 1981, The logic of multispectral classification and mapping of land. *Remote Sensing of Environment*, **11**, 231–245.

Rock, B. N., Moshizaki, T. and Miller, J. R., 1988, Comparison of *in situ* and airborne spectral measurements of the blue shift associated with forest decline. *Remote Sensing of Environment*, **24**, 109–127.

Rogers, R. H., 1973, *Investigation of Techniques for Correcting ERTS Data for Solar and Atmospheric Effects*, Technical Memorandum (Greenbelt, Maryland: NASA).

Rollin, E. M., Steven, M. D. and Mather, P. M. (editors), 1985, *Atmospheric Corrections for Remote Sensing* (Nottingham: Remote Sensing Society).

Ross, Yu. K., 1975, *Radiative Regime and Architectonics of Vegetation Cover* (Leningrad: Gidrometeoizdat).

Russell, E. W., 1973, *Soil Conditions and Plant Growth*, 10th edition (London: Longman).

Ryerson, R. A. and Curran, P. J., 1990, Agriculture. In *Manual of Photographic Interpretation*, 2nd edition edited by W. Philipson (Falls Church, Virginia: American Society for Photogrammetry and Remote Sensing) (in press).

Rytov, S. M., Kravtsov, Yu. A. and Tatarsky, V. I., 1978, *Introduction to Statistical Radiophysics. Part 2. Random Fields* (Moscow: Nauka).

Sadovnikov, Yu. N. and Orlov, D. S., 1978, Spectrophotometry to characterize soils, soil colouration and quantitative laws of reflection of the light by soils. *Agrochemistry*, **4**, 133–151.

Schubert, J., Chagarlamudi, P. and Jack, A., 1980, Stratification of Landsat data by productivity of soils. In *Machine Processing of Remotely-Sensed Data Symposium* (West Lafayette: Laboratory for the Applications of Remote Sensing), pp 186–194.

Sharma, R. C. and Bhargava, G. P., 1988, Landsat imagery for mapping saline soils and wet lands in north-west India. *International Journal of Remote Sensing*, **9**, 39–44.

Sharonov, V. V., 1958, *Nature of Planets* (Moscow: Fitzmatgiz).

Slater, P. N., 1980, *Remote Sensing: Optics and Optical Systems* (Massachusetts: Addison–Wesley).

Smelov, V. V., 1978, *Lectures on the Theory of Neutron Transfer* (Moscow: Atomizdat).

Smirnov, V. M., 1972, *Recommendations on Mapping Field Weediness and Measures Against Weeds* (Moscow; Saratov).

Sorokina, N. P., 1967, A quantitative estimation of the colour of typical chernozem. *Soil Science Institute Bulletin*, **1**, 116–125.

Spoor, G., 1975, Fundamental aspects of cultivation. In *Soil Physical Conditions and Crop Production* (London: HMSO) MAFF Technical Bulletin 29, pp 128–144.

Steglik, O., 1982, A technique for retrieving soil erosion based on remote sensing data. *Studies of the Earth from Space*, **3**, 92–95.

Stoner, E. R. and Baumgardner, M. F., 1980, *Physiochemical, Site and Bidirectional Reflectance Factor Characteristics of Uniformly Moist Soils* (West Lafayette: Laboratory for the Applications of Remote Sensing) Technical Report 111679.

Stoner, E. R., Baumgardner, M. F., Biehl, L. L. and Robinson, B. F., 1980, *Atlas of Soil Reflectance Properties* (West Lafayette: Purdue University, Agricultural Experiment Station), Research Bulletin 962.

Switzer, P., Kowalik, W. S. and Lyon, R. J. P., 1981, Estimation of atmospheric path-radiance by the covariance matrix method. *Photogrammetric Engineering and Remote Sensing*, **47**, 1469–1476.

Talmage, D. A. and Curran, P. J., 1986, Remote sensing using partially polarised light. *International Journal of Remote Sensing*, **7**, 47–64.

Tarasov, K. I., 1968, *Spectral Instruments* (Leningrad: Mashinostroenie).

Tarchevsky, I. A., 1977, *Fundamentals of Photosynthesis* (Moscow: High School Press).

Tatarinova, N. Ya., Kozlov, G. E. and Beliaev, V. A., 1980, *Measures Against Weeds in the Non-Chernozem Zone* (Moscow: Rosselkhozizdat).

Teillet, P. M., 1986, Image correction for radiometric effects in remote sensing. *International Journal of Remote Sensing*, **7**, 1637–1651.

Tikhonov, A. N. and Arsenin, V. Ya., 1979, *Techniques to Solve Incorrect Problems* (Moscow: Nauka).

Tiurin, I. V. and Konova, M. M., 1962, Humus biology and soil fertility. *Proceedings, International Conference on Soil Science* (Moscow: Sarakov), pp 67–74.

Tolchelnikov, Yu. S., 1974, *Optical Properties of Landscape* (Leningrad: Nauka).

Trudgill, S. T., 1988, *Soil and Vegetation Systems*, 2nd edition (Oxford: Oxford University Press).

Tulikov, A. M., 1974, *Techniques for Recording and Mapping Field Weediness* (Moscow: Saratov).

Vaksman, S. A., 1937, *Humus. The Origin, Chemical Composition and its Role in Nature* (Moscow: Selkhozgiz).

van den Bosch, R. and Messenger, P. S., 1973, *Biological Control* (Leighton Buzzard: Intext).

Vincent, R. K., 1973, An ERTS multispectral scanner experiment for mapping iron compounds. In *Proceedings, 8th International Symposium on Remote Sensing of Environment* (Ann Arbor: University of Michigan), pp 1239–1247.

Wardley, N. W., Milton, E. J. and Hill, C. T., 1987, Remote sensing of structurally complex seminatural vegetation—an example from heathland. *International Journal of Remote Sensing*, **8**, 31–42.

White, L. P., 1977, *Aerial Photography and Remote Sensing for Soil Survey* (Oxford: Clarendon).

Wood, T. F. and Foody, G. M., 1989, Analysis and representation of vegetation continua from Landsat Thematic Mapper data for lowland heaths. *International Journal of Remote Sensing*, **10**, 181–191.

Wooley, J. T., 1971, Reflectance and transmittance of light by leaves. *Plant Physiology*, **47**, 656–662.

Wright, W. D., 1969, *The Measurement of Colour* (London: Adam Hilger).

Zyrin, N. G. and Kuliev, F. S., 1967, Spectral reflectance of soils of the Lenkoran zone of Azerbaidjian SSR. *Scientific Papers of High School Biology*, **7**, 123–129.

Index

Page numbers in italics refer to Figures, Tables and Equations.

acetic acid 78
Acroptilon repens 134
adding functions *see* colour coordinates, specific
aerosols, atmospheric 151, *153–4*, 156, 159
aerosol size-distribution model 154, 156
Agricultural Mechanical Engineering, All-Union Research Institute for 140
Agricultural Meteorology, Institute for (Obninsk) 114
aircraft ix, 74, *127*
  atmospheric correction 151, *156*
  crop weeds 140, 147
  soil/crop state sensing *169–73*
  soil humus *99–102*
albedo *see* reflectance coefficient
alcohol 83, 117, 119
algorithm 142–3, 165
alkali 83
altitude 128, 140, 148, 170
  atmospheric correction *151–2, 154–5*
  surface reflectance 9, *12–16*
aluminium 75
  compounds 78, 82
aluminosilicates 79
anisotropy
  angular 9, 55
  angular-reflection 9, 12, 152, 154
  coefficient 4, 9, *12–15*
  reflectance *156*
approximation coefficients 156
atmosphere
  intensity (noise) 157
  optical properties 156, 168
  surface *167*
  transparent *164, 166*
  turbid 162, *164–6*
atmospheric correction model 157
atmospheric correction of remotely sensed data *151–68*
  influence of atmosphere *151–6*
  soil/crop state *174–5*
  techniques *157–68*
atmospheric haze 157
atmospheric radiation transfer model 157
atmospheric stratification model 163
atmospheric transfer function 151, 154, 156, 157
atmospheric transmission radiance model 168
atmospheric turbidity model 165
autocorrelation coefficient *60–1*

barley, 117, *119–22, 124–5, 138–40*, 146, *176*
Bayesian maximum likelihood 140
biogeographical survey 137
Boethe-Solpeter's equation 105, *107*
boundary-value problems 152, *157–62, 165*
brightness *see* intensity
brightness coefficient *see* reflectance
buckwheat 117, *119–21*
Burre approximation *107, 111–12*

calcite 78
capillar water capacity 80–1, 82

Index

carbon 83, 87
  organic 88
carbonate 74, 76, 78–9, 87
  calcium 82
cation exchange capacity 87–8
Central Institute for Agrochemical
  Service (CIAS) 101–2
Chebyshev's polynomials 100, 163, *164*
Chenopodium album 139
chlorophyll ix, 114
  concentration 146
  estimation by chemical analysis *116*,
    117–19, *121–3*
  estimation in laboratory and field
    115–125
  influence of leaf structure *117–22*
  extraction *117–22*
  index 122
  remote sensing of *115–25*
    and crop yield *122–5*
  leaf structure *117–22*
CIAS (Central Institute for Agrochemical
  Service) 101–2
CIE (Commission International de
  l'Eclairage) 21, *31*
classification
  crop *176–7*, *179–80*
  image 19, 167, *170–3*
  land cover ix, 18
  soil 86, 88
clover, red 117, *120–1*
coherence (correlation) function 105, *107*,
  109, *110*
colorimetric observer, standard
  (SCO) *26–30*, 38
colorimetry
  calculations 26, 29, 38
  definitions 19–20
  quantitative 117, 126, 133
  standards *26–31*
  systems *21–4*
  techniques 126, 169–70
Colorimetry, Committee on 20
colour ix
  characteristics 19, 24, *26–7*
  'linearity laws' 20
  measurement ix, *19–39*, 79
    calculations 21, *23–6*, *35–9*
    calculation techniques 24–6
    colorimetric systems *21–4*
    colorimetry standards *26–31*
    coordinates *31–5*
    definitions 19–20
    mixing 20–1

  quantification 21, 126–7, 131, 132,
    169–70
  saturation (purity) 19
  triangle *21–3*, 126
colour coordinates *21–3*
  from airborne measurements *171*
  calculations 26, 29, *35–9*
  crop weeds *144–5*
  humus *90–3*, 95
  soil/crop state *126–33*, 170
  techniques *24–6*
  and chlorophyll concentration *116–25*
  from field measurements *31–5*, 98–9,
    127–8, *130–1*
  humus *90–102*, 170
  from laboratory measurements *31–5*,
    91–8, *116*, 117–19, *121*
  soil/crop state *170–1*
  specific (adding functions) *24–9*, 38
  standards *27–8*
colority coordinates 21–3
  calculation techniques *24–6*
  crop state 126
  from laboratory measurements *33*
  soils 34
coltsfoot *139*
Commission International de l'Eclairage
  (CIE) 21, *31*
Copyright Agency (USSR) x
cornflower *139*
correction, radiometric 18, 151–2, 155,
  175 *see also* atmospheric
  correction
correlation coefficient *61*
correlation (coherence) function 105, *107*,
  109, *110*
cress, wild winter *138–9*
crop(s) ix, 99–100
  blossoming stage *179*
  bushing *179*
  calendar 176
  canopy 127, 131, 138, 172, 177
  cereal 176, *179*
  classification *176–7*, *179–80*
  colour 126–7, 176, *179*
    coordinates *36–9*, *126–32*
  damaged *126–7*, *129*, 130–1, 134
  earing-blossoming phase 138
  earing phase 124, *137–45*, 175, *176*,
    179
  green *175*, *179*
  growth 134, 135
  harvesting 134, 135, 144–5
  healthy *126–30*

192

Index

height 146–7, 175–6, 179
milk-ripeness stage *176*
pests and diseases 134
phenophases *175–6, 179*
plant-emergence phase 175
quality 134
reflectance *127*, 128–9, 138, 173, *175–7, 179*
ripening 176, *179*
rotation 135
senescence ix, 126, 175
sowing 135, 148
spring *175–9*
sprouting *179*
stalk-growth phase 175
state (condition), remote sensing of *126–33*
  sensing of *125–33*
  aerovisual estimates *126–31*
  field estimates *126–8, 130–1*
  laboratory calculation *127*
stubble *179*
tubing stage *176, 179*
types 179–80
unhealthy *127–31, 176, 179*
wax-ripeness stage *145–8, 176*
winter *126–33, 175–80*
yellow *175, 179*
yield *122–5*, 126, 134
see also chlorophyll; weed(s)

deciduous canopies 172
desert *83–4, 152, 154–6*
diffraction
  light 6, 8
  radiation 43, 44, *55–8*, 106
  small perturbations 43
  tangent-plane 43, 51, *54–5*
  wave 49, *55–8*, 106
  see also scattering
diffuser 17
disturbance theory *107*, 108, *111*
Dyson's integral equation 105, *107*

ecology 115
eigenfunctions *111*
eigenvalues *111*
eigenvector *108*
electrodynamics, quantum 105
elevation, solar 124, 128, 140, 151
  soil reflectance 74, 99–100
  surface reflectance 9, *12–17*
erodability 87–8, 169, 172
erosion 74, 87–8

Feinman's diagrams 105, 106
feldspar 78
fertilizer 138
field analysis, traditional *130–1, 132–3, 144–5*
filtering 166–7
fire 135
flux, light 17, 22
flux, radiation 45
flux, radiative *4–5*, 20, 24, 26, 31
forest (taiga) *9–12, 15*, 17, *172*
  colour measurement 29, *32–4*, 126
  see also soil(s)
Fourier series 51, *53, 158*
Fourier transformation *59, 109, 111–12*, 166
Fraunhofer zone 50, *54*
Fresnel formulae 43–4
Fresnel reflection 7
fulvic acid 75–6, *83–6*
function of errors *142*

gases 151, 159
Gaussian distribution *60–1*, 105–6
geographic information systems ix
geographical referencing 148
geology 173
geometric optics *43–9*, 55
geometry 98, 105, 138
  plane-parallel 162–3
  soil reflectance 43, 80
  view 3, 9, *31*
grass 12, 17, *172, 175–6*
Green's formula *49–50*
Green's functions *50*, 105, *107–9*
gypsum 78

Henye-Greenstein indicatrix *161*
herbicides 135
histogram *171–2, 177*, 179
humic acid 75–6, *83–6*
humins 83
humus, soil ix, 18
  colour measurement *32–4*
  reflectance 64, 74–6, *77*, 78–9, *80–2*
  remote sensing of *83–102*
    aircraft studies *99–102*
    chemical composition and analysis *83–6*, 91, *94*, 97
    field studies 86, 98–9, *100*, 102
    laboratory studies 86–7, 91, *94–100*, 102
    reflectance properties 86–9
    theoretical studies *89–91*

hydration 76–7, 82
hydroxyl 88

indicatrix of radiation 3–4, *5*, 6–7,
   9–12, 17
intensity (brightness) *3–4*, 6, *10–12*, 90
   atmospheric correction 152, 154, 157,
      162–6
   crop state sensing 173, *175–7*
   crop weeds 141, 148
   soil reflectance theory 45, 47–8,
      59–61, 79
   vegetation reflectance *110*, *112*, 114
International Luminance Council (ILC)
   21, *24–30*
iron 78, 87–8, 172
   compounds 76–7, 78
   oxide 74–9, 82, 86, 87–8
      ferric and ferrous 80
   peroxide 76
   protoxide 76, 78
   silicate 77

kaolinite 78

Lambert's law 45
land cover ix, 18, *178*, *179*
land potential 169
land use 169
Laplace operators *107*
leaching 77
leaf
   chlorophyll concentration *115–17*
   colour 126
   potato *116–17*
   reflectance 115, *117–22*
   structure, species-specific *117–22*
Legandre polynomials *159*, *161*, 165
Leningrad State University atmospheric
   model *153*, 156
linear systems theory 165
'lobes' *see* spectral harmonics
loess *see* soil(s); soil-forming rocks
LOWTRAN (atmospheric transmission
   radiance model) 168

Main Geophysical Observatory (MGO)
   151
maize *176*
Maltsev scale 136–7
manure 135
map, thematic *178 see also* soil(s)
Mapper, Thematic (TM) 87, 88
Marshak form *159*, *161*

matrix 142–3, *160–2*
   covariance *141*
   unit *142*
mean-square deviation (MSD) 146
mean-statistical model 163
Meteor (MSU-M) *151–6*, 167, 169–70,
   174–5, *177–80*
Mie theory 154
milk glass, MS-13 140
mineral acids 83
minerals 78, 134, 135
   silicate 77
modelling
   atmospheric *153–4*
      aerosol 154, 156
      correction 157
      mean-statistical 163
      radiation transfer 157
      stratification 163
      transmission radiance 168
      turbidity 165
   physiomathematical 105
   radiation/soil-crop system 180
   refractive index 154
   spectral mixture 129, 133
   vegetation canopy reflectance *105–14*
monochromatic signals 140
Moscow State University-type Meteor
   (MSU-M) *151–6*, 167, 169–70,
   174–5, *177–80*
multispectral data 151, *154–6*
Multispectral Scanning System (MSS)
   88, 167

NASA (National Aeronautics and Space
   Administration) x, 169, 180
National Research Council (USA) x
neutron transfer theory 158–9
nitrogen 83

oats *138–40*, 146
optical density 17, 88, *170–3*, 175–6,
   177, *179–80*
optical theorem *112*, 162
ordinates *24–6*, 100, 122
organic matter/materials 75, 76, 78, 82,
   87–8, 135
oscillogram 100
ox-eye daisy *138–9*
oxidation 76, 82
oxygen 83

photodensitometer, P-1000 170, 172
photoelectric measurement 17

Index

photographic measurement 17
photography, multispectral aerial 88
photometer 17, 114, 124, 128, 132, 138, 146–7
photometric wedge 151–2
photomultiplier, FEU-62 140
photosynthesis 135
physics, nuclear 157
physiology, plant 115
phytomass 173–6, 180
phytosociological survey 137
pixels 152, 165–7, *171*, *177*
Planck's law 26
polarisation 43
potato *116–17*, *120–1*
precipitation *83–4*, 87, 99, 148, 173
principal component analysis 88
probability density *56–7*, *60*

quartz 77–8
  ionized 76

radar, synthetic aperture ix
radiance 4–5, 7, 59–50, 141
  in atmospheric correction 155, *159–60*, 162–3, 165
  beam *108*, *110*
  diffuse *158*
  model, atmospheric transmission 168
  surface *158*
  vegetation canopy reflectance *107–12*
radiation 5–6, 173
  absorption *81*, 89
  angular distribution *6–8*, *47–8*, *61*, 75, 151
  atmospheric transfer model 157
  backscattering 111
  chlorophyll reflectance 124
  colorimetry 26, 38
  colour coordinates *24–6*, 31
  colour mixing 20
  diffraction 43, 44, 49, *55–8*, 106
  diffuse *7–8*
  direct 99
  directional *112*
  electromagnetic ix
  field *105–8*, *111–12*, 173
    laws of formation 157
    theory 105, 114
  global 7, 31, 44, *112*
  incident *24–6*, 31, 43, 44, 89
  indicatrix *3–4*, *5*, *6–7*, *9–12*, 17
  interaction with vegetation 105, 108
  reflected 7, 140

reflection *43–61*
  geometric optics *43–9*, 55
  tangent plane technique 43, *49–61*
    sinusoids *52–6*
    statistically rough surface *56–61*
  scattered *44–9*, 56, 89, 99, 106
  soil-crop system model 180
  solar 6–8, 99
    in soil reflectance theory 44, *47–8*, *56–9*, *61*, 79–80, 82
  standard *25*, 127
  transfer 151, 157
    equation *108–12*
    theory 43, 49, 105, 108, 111, 157
  transformation 151
radiometer 72, 99, *101*, 151–2, *153*, 155
radiometry 175
radish, wild *138–9*
random continuous medium 106, 114
Rayleigh atmospheric model *153–4*, *156*
reflectance (brightness coefficient) x, 3, 5–6
  atmospheric correction 151, *154–7*
  calculations 112, 114, *145–8*
  coefficient *3–4*, 5–7, 115
    atmospheric correction 151–2, 154–5, *156*, *161–2*, *164–5*, *167*
    humus 86, *89*
    soils 75–8, *79–80*, 82
  indicatrix 151
  measurement 90, 138, 144–6
    airborne *8–9*, 34–5
    atmospheric correction 151–2, *156*
    crop state 128–9, 132
    crop weeds 140, 147–8
    humus *99–102*
    in field 17, 34–5, 62
      crop state *127–8*, 131–2
      soils *72–4*
      vegetation *112–14*
    instruments 17
    in laboratory 17, 34–5, 62, 63–4, *72–4*, 124
    spaceborne 151, 155, *156*, 168
    techniques 17
      Fresnel formulae 43–4
      geometric optics *43–9*, 55
      radiation diffraction 43, 44, 49
      radiation transfer 43, 49
  natural surfaces *3–12*
  orthotropic (Lambertian) *6–7*, 30, 46, 55
profiles, temporal 180
specular 6–7, 15, 44, 55, *57–61*, 80

see also crop(s); humus; leaf; soil reflectance; surface(s); vegetation reflectance
reflection
  angular distribution measurements *8–17*
    indicatrix of radiation *9–12*
    -anistropy coefficient *12–15*
    -asymmetry coefficient 9, *13–17*
  coefficient 75–6, 86–7, 115
  indicatrix *161*
  Lambertian *161*
  non-orthotropic *161*, 163, 165
  see also radiation reflection
refractive index model 154
remote sensing systems 17–18, 88, 140
remote sensing techniques and instruments 86, 138, 146, 169
remotely sensed data 18, 74, 76, 82, 169
  crop state 127–8, 130–1, 169, 179
  digitised 157
  humus 87, 88–90, 95, 102
  vegetation 106, 114
  weediness 137–8, *146–8*
  see also atmospheric correction
remotely sensed imagery 167, 175
remotely sensed measurement 111, 114, *142–4, 154–6*
reservoirs *177–9*
Royal Society (UK) x
rye *138–40, 142–5*

salinity 74
salt, soluble 74, 78
satellite, Meteor MSU-M *151–6*, 167, 169–70, 174–5, *177–80*
satellite sensor ix, 74, 102, 151, 157, 165–8
  soil/crop state assessing *169–75*, 179–80
scatterers 3, 6–8, 44, 106–7, *108*, 114
scattering *50, 110, 112*, 154
  macroscopic *159*
  multiple 43, 49, 55, 105–6, 108, 114, 167
  single 43
sciences, physical 151
semivariograms 167
sensor imagery 165, 169–70, *171*, 173, *175–80*
sensors *4*, 17, 31, 98, 105, 111, 146
  airborne 34–5, 74, *99–102*, 151–2, *156, 169–73*

scanning 162–3
spaceborne 74, 151, *156–7*, 165–7, *169–75, 177–80*
  Thematic Mapper (TM) 87, 88
shepherd's bag *138–9*
silicon 75, 82
sinusoids 49, *52–6*
small perturbations technique 43
snow *174–5*
soil(s) ix
  aggregates *79–80*, 82, 99
  air-dried 63, 64, 81, 91
  alfisol 87
  arable 34, 63, *136*
  background 124, 127–8
  bare *127–8*, 130, *132*, 135
  boggy 76
  brown 63–4, *76–7, 80*, 81, *82, 84*
  chemical composition 169
  chernozem *84*, 86, 138, *173–4*
    colour measurement *32–4, 36–7*
    reflectance 63–4, *65, 68–9, 73–7, 80–2*
  classification 86, 88
  clay 87–8
  colour 72, 74–9, 82, 87, 169–70
    coordinates *31–9, 90–102, 127–9, 132*
  compressed *90*
  cover *113–14*, 131–2, 141, *175–6*, 180
  desert *83–4*
  fallow *175*
  fertility 86
  forest 34, 63–4, *65, 73–6*, 78, *82, 84*
  formation 86
  gley 64, *65*, 76
  harrowed 44, *46–9*, 55, *58*, 135
  heat balance 74
  horizons *63–5*, 72, 76, *79–80*, 90
  lateritic *84*
  loess *70–2, 90–102*
  mapping 87, 88, 99–100, *101*
  microclimate 169
  moisture ix, 34, 124, 135
    in humus 87–8, 98–100
    and reflectance 63, 72, 74, 77, 79–81, *82*
    in soil state sensing 169, 172–4
  mollisol-type 87
  morphology 74
  oven-dried 81
  peat-podzol *32–4, 84*
    reflectance 64, *65–7, 73–5*, 78, *80–2*

Index

ploughed 7, 63, 74, 135
    reflectance theory *43–9*, 55, *57*
    and remote sensing 169–70, 172–5, *178–9*
        of humus *90*, 98, 99–100
    properties 72, 88, 89, 169
        chemical 74–9, 82, 83
        physical 74, *79–82*, 83
        physicochemical 87
    red 77, *80–2*, *84*
    reflectance measurement instruments 17
    state (condition), remote sensing of *173–80*
        from aircraft and satellites *169–73*
        laboratory analysis 137
    roughness 63, 79
    saline *80*, *82*
    sandy 7, *65*, 87–8, *172*
    shaded *49–52*, 112, 114
    steppe *83–4*
    structure 34, 74, 82
    subtropical 76, *77*, 78
    texture 74
    topsoils 63–4, *65–9*
    type 18, 34, 172
        and reflectance 63, 74, 76, *77*, 79, 81
        and remote sensing of humus *85*, 87, 90
    uncultivated *44–6*, *55–6*, 135
    utisol 87
    yellow *76–7*, *80–2*
    see also humus
soil-crop system model 180
soil-forming rocks 63, *70–2*, 77, *89–102*
soil reflectance
    atmospheric correction 151, *174–5*
    in laboratory and field *63–82*, 170
        chemical properties *74–9*, 82
        field measurements *72–5*
        by horizon *64–5*
        physical properties 74, *79–82*
        soil-forming rocks *70–2*
        topsoils *63–9*
    and soil/crop state 169–70, *172–4*, 176
    theory *43–62*
        radiation-soil interaction *44–9*
        rough surface *49–61*
            sinusoids *52–6*
            statistically *56–61*
    see also humus
Sonchus arvensis *139*, 146
sorgo 117, *120–1*

soybean 115
spectral brightness *see* spectral intensity
spectral brightness coefficient (SBC) 4, 6–7, 9, 17
    and colour 26, 31, *35–6*, 38
    crop chlorophyll 122, 124
    crop weeds *138–9*, 140–4, *145–7*
    humus 87–8, 90, *95–6*, 98–100
    soil/crop state 127–8, 132, 170, 173
    soil reflectance 63, 72, 74
    in soil reflectance theory *44–6*, *48*
spectral channels 151–2, *153*, 155, 167, *174*
spectral difference density *109–10*
spectral harmonics ('lobes') 53, *54*, 55–6, 61
spectral intensity (spectral brightness) 4, 9, 12, 17, 19, 72, 100, 179
spectral intervals 124, 173
    atmospheric correction 152, 154, *158–9*, 162
    colour 24–5, 35
        coordinates *36–9*
    crop weeds 141, 143, 145
    in humus 88, 91
    natural surfaces 9, 12, 15, 17
    in soils 76–7, *81*
spectral mixture 128, 132
    model 129, 133
spectral radiance 127
spectral ratio *140*
spectral-reflectance coefficient (SRC) 4, 6
    chlorophyll 115, 118, *120*, 122
    and colour 24, 26, *29–30*, 35
    of humus 90, 98
    of soils *63–73*, 80–1
spectral-reflectance curve *35–6*, 112–14
    chlorophyll 115, 117, 119, *121–2*, 124
    crop weeds 138, 146
    humus 86–8, 91, 95, 99–100
    soil/crop state *127–9*, 131, 170, 175
    soils 63, 64–5, 77, 80, 82
spectral shift 125
spectrogram 100
spectrometer
    fast-operating 100, *170–1*
    field 17, 74, 79, 98
    ground-based *174*
spectrometry ix, 99
    principles *3–18*
        distribution of radiation *6–8*
        distribution of reflection *8–9*
        indicatrix of radiation *9–12*

parameters and terms *3-6*
reflection-anisotropy coefficient *12-17*
techniques and instruments 17
spectrophotometer, SF-18 17, *34-6*
  crop chlorophyll 117-19, 122
  remote sensing of humus 91, 95, 99
  soil reflectance 64, 72
spectrophotometric coefficients 78
spectrophotometry 79, 99, 148
spectrum, electromagnetic 63
spherical harmonics *159-61*
standard colorimetric observer (SCO) *26-30*, 38
steppe *83-4*
sun
  atmospheric correction 152, *154-5*, 157
  soil reflectance theory 46, *48*
  surface reflectance *7-9*
  vertical plane *8-11*, *13-17*, *48*, 152
sunflower *176*
surface(s)
  atmosphere *167*
  colour measurement *29-30*
  cultivation 172
  desert *154-6*
  Earth 151, 157, 161, 167
  grass 12, 17
  intensity 157, *163-4*, *166*
  natural *3-18*
    distribution of radiation *6-8*
    mixed *6-8*
    reflectance measurement techniques 17
    reflectance parameters *3-6*
    reflectance properties *4*, *9*
    spectral properties *17-18*
    vegetation reflection measurements *8-17*
  orthotropic (Lambertian) *6-8*, 162, *164*
  radiance *158*
  reflectance *165*, 167, 174
  reflecting 43
  ridge and furrow 44, *45-7*, *57*
  rough *6-8*, 106, *107-8*, *111-14*
    see also soil reflectance
  scattering 50, *58-9*
  sinusoidal *61*
  smooth 43-4, *45*, *53*, *55-6*, 61, 111
  specular *6-8*, *164*
  spheroid aggregate 44, *46-8*, *58*
  statistically rough *56-61*

structure 44

tangent-plane technique 43, *49-61*
  sinusoids *52-6*
  statistically rough surface *56-61*
temperature *83-4*, 87
Thematic Mapper (TM) 87, 88
tillage 169
topography 157, 167, 173
turbidity, atmospheric *165-7*
  model 165

ultraviolet 83, *85-6*

vector(s)
  atmospheric correction *159-62*
  crop weeds *141-4*
  -matrix *160-2*
  orthogonal *160*
  in soil reflectance *49-52*
  vegetation canopy reflectance *108-10*, *112*, 114
vegetation ix, 83
  canopy *105-14*, 127, 130-1
  colour *127*
    coordinates *35-9*
  cover 127-8, 173, *179*
  properties 106
  reflectance 43-4, 117, 125,
    canopy modelling *105-14*
      field evaluation *112-14*
      three approaches *106-13*
      USSR approaches 105-6
  remote sensing of 133, 173
  surface reflectance *7-18*, *174-5*
  state (condition) 114, 127
viewing angles 9, 12, *30-1*, 146-7
  in atmospheric correction 152, *154-5*, 158-9
  in soil reflectance theory *44-8*, *55-6*, 60

wave(s)
  diffraction 49, *55-8*, 106
  electromagnetic 43, 49, 114
  equation 105, 108, *111*
  Huggens 50-1, *59*
  incident 53, *55*, 114
  monochromatic 49
  numbers *52-4*, *107*, *111*
  plane 50, *111*
  plane monochromatic 49
  scattered 49-50, *52-5*, *59*
  spherical 50, *108*

vectors *52–3*
waveguide *111–12*
wavelength
  atmospheric correction 151, *153*, 155
  and colour 21, *23*, *27–30*, 35–6
  crop chlorophyll 115, *117–20*, 125
  crop state *127*, *174–7*
  crop weeds *139–40*, *145*, 146
  humus 83, *85*, 86–9, *95*, 100
  in soil reflectance 43, 55, 63, *64–5*, 75, 77–8, *80–1*, 82
  surface reflectance 9, *12–17*
  vegetation canopy *107*, 111, *112–14*
weed(s) ix, 114
  biological composition 137
  canopy 138, 144
  concentration 134, 142–3
  control measures 135, 137, 148
  crop, remote sensing of *134–47*
    classification 134–5, 136, 138 *140–1*, *143–4*
    control 135, 137, 148
    earing phase *137–45*
    recording *135–7*
    wax-ripeness stage *145–8*
  estimation
    quantitative (weight) 136, 137
    by remote sensing *137–48*
    visual (qualitative) 136–7
  feeding characteristics 134–5
  fertility 135
  germination 135
  height 146–7
  identification 138
  map 135–8, *147–8*
  multi-year 134, *136–7*
  non-parasitic 134–5
  one-year 134, *136–7*
  parasitic 134–5
  propagation 134
  reflectance *138–9*, 147
  seeds 134–5, *136*
  semiparasitic 135
  spring 134
  survey 140
  winter 134
  wintering 134
wheat *138–40*, *145*
  winter 134, 175, *176*, 180

Author index

Adams, J. B. *et al.* 128-9, 141
Ahern, F. J. *et al.* 18
Akhmanov, S. A. *et al.* 114
Allen, W. A. 127
Antokolsky, M. L. 49, 50-1
Asmus, V. V. *et al.* 52, 157
Arsenin, V. Ya. 166

Barahanenkov, Yu. N. 111
Baslavskaya, S. S. 117, 119
Bass, F. G. 19, 105, 106, 107-8, 111, 112
Baumgardner, M. F. 87
Beckett, P. H. T. 82
Benedict, H. M. 115
Bhargava, G. P. 74
Borisoglebsky, G. I. 170, 173, 175
Born, M. 44, 106, 162
Bouma, P. J. 20, 26
Bowers, S. A. 74
Brady, N. C. 44, 75, 76-7, 83, 169
Brandt, A. B. 19, 115
Brekhovskikh, L. M. 43, 49, 50, 51, 52-3, 55
Bridges, E. M. 74
Briggs, D. J. 135
Brisco, B. *et al.* 173, 176
Butkovsky, A. G. 17, 165

Canosa, J. A. 157
Chamberlin, G. J. 20
Chaume, D. 88
Chewings, V. H. 74
Colwell, J. E. 9
Condit, H. R. 35
Courtney, F. M. 135

Curran, P. J. 17, 74, 92-3, 99, 114, 115, 133, 151, 167
Curtis, L. F. *et al.* 76

Dacosta, L. H. 88
Dave, J. V. 157
Dawson, J. H. 134
Deering, D. W. 3-4
Dergacheva, M. I. 83, 86
Dirmhirn, I. 6
Drury, S. A. 19
Duggin, M. J. 4
Dzhad, D. 20, 26, 30, 127

Eaton, F. D. 6
Eck, T. F. 3-4
Ellenberg, H. 137
Escadafal, R. *et al.* 19, 24, 26
Evans, R. 74, 88
Evans, S. A. 135

Fedchenko, P. P. ix, 7, 17, 35, 43, 63, 74, 90-1, 99, 114, 127, 128, 138, 140, 143, 146-7, 169, 170, 173, 179
Feldbaum, A. A. 17, 165
Fisiunov, A. V. 134, 136
Foody, G. M. 141
Foody, G. M. *et al.* 176
Forster, B. C. 18
Fryer, J. D. 135
Fuks, I. M. 19, 105, 106, 107-8, 111, 112

Gausman, H. W. 127
Gazarian, Yu. L. 111
Goncharsky, A. V. *et al.* 166

Author index

Goudie, A. 169
Griggs, M. 157
Grishina, L. A. 86
Gurevich, M. M. 21

Hanks, R. J. 74
Harold, R. W. 24, 26
Hay, A. M. 99
Holstun, J. T. 134
Horlet, D. N. H. 125
Horwitz, W. 115
Hunt, R. W. G. 21, 22, 23, 26
Hunter, R. S. 24, 26

Isakovich, M. A. 43, 49, 50
Isimaru, A. 105, 106–7, 112, 114

Johanssen, C. J. 88
Judd, D. B. 21, 24, 26, 30, 31

Karmanov, I. I. 43, 63
Kastrov, B. G. 46–7
Kauli, J. 106
Kelly, K. L. 21
Khabibrakhmanov, Kh. 137
Kharin, N. G. 35
Kiefer, R. W. 17
Kimes, D. S. *et al.* 18, 30
Kirchner, J. A. 30
Kizel, V. A. 44
Kliatskin, V. I. 114
Kneizys, F. X. 168
Koltsov, V. V. 17, 74, 98, 100, 170
Kondratyev, K. Ya. ix, 7, 17, 35, 43, 63, 74, 90–1, 99, 114, 127, 128, 138, 140, 143, 146–7, 169, 170, 173, 179
Kondratyev, K. Ya. *et al.* 7, 9, 17, 21, 60, 90–1, 117, 119, 122, 154, 157, 180
Konova, M. M. 83, 84
Korzov, V. I. 9, 12, 15, 63
Kozoderov, V. V. ix, 105, 157, 161, 165, 170, 173, 175
Kozoderov, V. V. *et al.* 6, 7, 161, 167
Krasilschchikov, L. B. 9
Kuchler, A. W. 137
Kulebakin, V. S. 45
Kuliev, F. S. 43, 63, 86, 88, 169
Kuz'mina, E. F. 83, 86

Latz, K. *et al.* 87, 88
Le Grand, Y. 26
Leshkevich, G. A. 30

Li, R. Y. *et al.* 18
Liberstein, I. I. 137
Lillesand, T. M. 17

MacAdam, D. L. 20, 21
Maltsev, A. I. 136
Mani, V. S. *et al.* 134
Markov, M. V. 136
Mather, P. M. 88
Matiukha, L. A. 136
Merik, B. 26
Messenger, P. S. 135
Mikhailova, N. A. 86–7, 88, 169
Milton, E. J. 17, 115
Mishin, I. V. 161, 165
Moissichik, V. A. 131, 133
Mueller-Dumbois, D. 137
Myers, V. I. 133

Neema, D. L. *et al.* 80
Nelson, D. J. 74
Newton, Sir Isaac 20
Nichiporovich, A. A. 122
Nikolsky, V. V. 111

Obukhov, A. I. 43, 63, 77–8, 79, 86, 169
Ogorodinkov, B. I. 126
Orlov, D. S. 43, 63, 74, 75, 76, 77–8, 79, 80, 81, 86, 169
Orlov, D. S. *et al.* 43, 63, 75, 76, 78, 79, 86

Panfilov, A. S. 6
Peacock, K. 157, 167, 168
Phu, T. N. 88
Pickup, G. 74
Pokrovsky, G. I. 86, 89, 169
Proshina, N. V. 43, 63, 75, 79

Rachkulik, V. I. 35, 63, 98, 173
Robinove, C. J. 19
Rock, B. N. 125
Rogers, R. H. 157, 167, 168
Rollin, E. M. *et al.* 18
Ross, Yu. K. 106
Russell, E. W. 83, 169
Ryerson, R. A. 133
Rytov, S. M. *et al.* 44, 105, 106, 109, 111, 114

Sadovnikov, Yu. N. 63, 74, 75, 80, 81
Sharma, R. C. 74
Sharonov, V. V. 6, 7, 44

Author index

Schubert, J. *et al.* 88
Shulgina, N. B 157
Sitnikova, M. V. 35, 63, 98, 173
Slater, P. N. 4, 105, 154
Smelov, V. V. 160
Smirnov, V. M. 136–7
Smoktiy, O. I. 157
Sorokina, N. P. 86–7, 88, 89
Spoor, G. 169
Steglik, O. 88
Stoner, E. R. 87
Sushkeich, T. A. 165
Swidler, R. 115
Switzer, P. *et al.* 157, 168

Tageeva, S. V. 19, 115
Talmage, D. A. 114
Tarasov, K. I. 17
Tarchevsky, I. A. 122
Tatarinova, N. Ya. *et al.* 135
Teillet, P. M. 18, 175
Ter-Markaryants, N. E. 9, 15, 63
Tikhonov, A. N. 166

Tiurin, I. V. 83, 84
Tolchelnikov, Yu. S. 35, 43, 63, 86, 88, 89, 169
Trubetskova, O. N. 117, 119
Trudgill, S. T. 83
Tulikov, A. M. 137

Vaksman, S. A. 83
van den Bosch, R. 135
Vincent, R. K. 76
Vyshetski, G. 20, 26, 30, 127

Wardley, N. W. *et al.* 5
Webster, R. 82
White, L. P. 74
Wolf, E. 44, 106, 162
Wood, T. F. 141
Wooley, J. T. 117
Wright, W. D. 21
Wyszecki, G. 21, 24, 26, 30, 31

Zonneveld, I. S. 137
Zyrin, N. G. 43, 63, 86, 88, 169